air conditioning service manual

Auto, Farm Equipment and Truck

2nd Edition

CONTAINS
- INTRODUCTION
- THEORY OF OPERATION
- UNIT APPLICATION
- SAFETY
- TROUBLESHOOTING
- MAINTENANCE
- REPAIR
- GLOSSARY

Published by

TECHNICAL PUBLICATIONS DIV.

INTERTEC PUBLISHING CORPORATION

P.O. Box 12901, Overland Park Kansas 66212

©Copyright 1985 by Intertec Publishing Corp. Printed in the United States of America.

All rights reserved. Reproduction or use, without express permission, of editorial or pictorial content, in any manner, is prohibited. No patent liability is assumed with respect to the use of the information contained herein. While every precaution has been taken in the preparation of this book, the publisher assumes no responsibility for errors or omissions. Neither is any liability assumed for damages resulting from use of the information contained herein. Publication of the servicing information in this manual does not imply approval of the manufacturers of the products covered.

All instructions and diagrams have been checked for accuracy and ease of application; however, success and safety in working with tools depend to a great extent upon individual accuracy, skill and caution. For this reason the publishers are not able to guarantee the result of any procedure contained herein. Nor can they assume responsibility for any damage to property or injury to persons occasioned from the procedures. Persons engaging in the procedures do so entirely at their own risk.

INDEX

	Paragraph
Abacus Compressor	
Checking Oil	68
Repair	94
ACCUMULATOR-DRIER	
APPLICATION	26
EXPLANATION	15
ATTACHING MANIFOLD AND LINES	47
Automatic Switches	19
CCOT . . . . Refer to Cycling Clutch Orifice Tube	
Charging (Filling) System	63
CHECKING OIL	65
Circuit Breakers	19
CLUTCH AND COMPRESSOR	80
COMPONENT TEST	54
Compressor Drive Clutch	
Application	21
Remove and Reinstall . . . . Refer to Specific Compressor	
Test	58
Compressor Front Head	
Delco Air A-6	92
Compressor Overhaul	
Delco Air A-6	86
Delco Air R-4	93
Tecumseh	104
York	110
Compressor Rear Head & Valves	
Delco Air A-6	98
Compressor Shaft Front Seal	
Delco Air A-6	84
Delco Air R-4	90
Sankyo-Abacus	97
Tecumseh	102
York	108
COMPRESSOR	
APPLICATION	20
EXPLANATION	7
CONDENSER	
APPLICATION	22
EXPLANATION	8
CONTROLS	
APPLICATION	19
EXPLANATION	17
Cycling Clutch Orifice Tube (CCOT)	
Clutch	21
Control	17
Orifice	11
Troubleshooting	57A
Delco Air A-6 Compressor	
Checking Oil	67
Repair	81
Delco Air R-4 Compressor	
Checking Oil	66
Repair	87
Detector Torch Leak Test	43
DISCHARGING SYSTEM	60
Dye Solution Leak Test	45

	Paragraph
EFFECTS OF PRESSURE	4
Electronic Detector Leak Test	44
EVACUATING SYSTEM	62
EVAPORATOR	
APPLICATION	25
EXPLANATION	12
EXPANSION ORIFICE (VALVE)	
APPLICATION	24
Control	17
EXPLANATION	9,10,11
Test	57
EXPLANATION OF BASIC PARTS	6
Fan or Blower Switch	19
FILLING (CHARGING) SYSTEM	63
FILTERS	16
Fixed Orifice	
Application	24
Explanation	11
Front Seal	
Delco Air A-6	84
Delco Air R-4	90
Sankyo-Abacus	97
Tecumseh	102
York	108
Fuses, Circuit Breakers & Automatic Switches	19
HEAT MOVEMENT	2
HOSES, TUBES & CONNECTIONS	14
INTRODUCTION	1
ISOLATING AND PURGING COMPRESSOR	64
Lack of Activity	38
LATENT HEAT	3
Leak Test Compressor	
Delco Air A-6	89
Delco Air R-4	91
Magnetic Clutch	
R&R Delco Air A-6	83
R&R Delco Air R-4	89
R&R Sankyo-Abacus	96
R&R Tecumseh	101
R&R York	107
Test	58
Main Control Switch	19
MAINTENANCE	59
Noise	39
Not Cooling	40
OPERATIONAL CHECK	37
ORIFICE	
APPLICATION	24
EXPLANATION	9,10,11
PRELIMINARY CHECKS	36

INDEX Continued

	Paragraph
R&R Clutch Assembly	
Delco Air A-6	83
Delco Air R-4	89
Sankyo-Abacus	96
Tecumseh	101
York	107
R&R Compressor and Clutch	
Delco Air A-6	82
Delco Air R-4	88
Sankyo-Abacus	95
Tecumseh	100
York	106
RECEIVER-DRIER	
APPLICATION	23
EXPLANATION	15
Test	56
RECYCLING REFRIGERANT	5
Reed Plate Assembly	
Delco Air A-6	86
Delco Air R-4	93
Sankyo-Abacus	98
Tecumseh	103
York	109
REFRIGERANT LEAK TEST	41
REFRIGERANT	13
REPAIR	79
ROUTINE SERVICE	59
Safety Switches (Controls)	
APPLICATION	21
EXPLANATION	17

	Paragraph
SAFETY	27
Sankyo-Abacus Compressor	
Checking Oil	68
Repair	94
Soap Solution Leak Test	42
Super Heat Switch	
APPLICATION	19
EXPLANATION	17
Testing	57
SWEEP-TEST CHARGE	61
SYSTEM PRESSURE TEST	48
Tecumseh Compressor	
Checking Oil	69
Repair	100
THEORY OF OPERATION	2
Thermostatic Control Switch	
APPLICATION	19,21
EXPLANATION	17
Test	55
TROUBLESHOOTING	35
Component Test	54
Leak Test	41
Pressure Test	48
UNIT APPLICATIONS	18
York Compressor	
Checking Oil	70
Repair	105

Technical Publications Div., is grateful to the following manufacturers for their wholehearted cooperation in supplying photos, drawings, technical information and dimensional data for this manual.

Allis-Chalmers Corp.
Ansel Manufacturing Co., Inc.
J. I. Case, Agricultural Equipment
   Div. TENNECO
Deere & Co.
Deutz Corp.
Eagle Air Conditioning
The Egging Co.

Ford Motor Co.
Frigidaire Div. of G.M.
Full Vision, Inc.
International Harvester Co.
K-D Manufacturing Co.
Massey Ferguson, Inc.
John E. Mitchell Co.
Nuday Co.

Royal Industries, Hinson Div.
Sankyo International
SGM Company, Inc.
Steiger Tractor, Inc.
Tecumseh Products Co.
White Motor Corp.
Year-A-Round Cab Corp.
York Div. of Borg Warner Corp.

This service manual provides specifications in both the U.S. Customary and Metric (SI) system of measurements. The first specification is given in the measuring system used during manufacture, while the second specification (given in parenthesis) is the converted measurement. For instance a specification of "0.011 inch (0.279 mm)" would indicate that the equipment was manufactured using the U.S. system of measurement and the metric equivalent of 0.011 inch is 0.279 mm.

INTRODUCTION

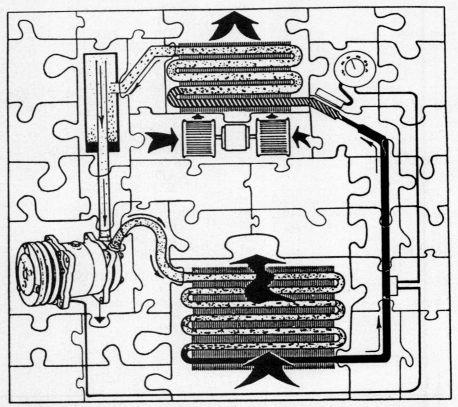

The method of cooling an object using a mechanical compressor and Refrigerant-12 is simple and should not remain a puzzle.

1. We have learned to appreciate temperature within a narrow "COMFORT RANGE". The most comfortable temperatures in degrees will depend upon humidity, altitude, what we are doing, our physical condition, and many other considerations; but we all find that we are able to do more and better work when the temperature is within a narrow range of perhaps six to 10 degrees. In certain situations the air surrounding our bodies must be heated or cooled to sustain life, but more often we warm or cool the air around us to be more comfortable.

We have learned to change and control the temperature of the air in our houses, businesses, automobiles, trucks, busses, trains, planes, and about everywhere that we can contain an environment including outer space. Controlling the quality of air (by filtering), humidity of air (by removing or adding moisture) and temperature of air (by heating and cooling) in our immediate environment is certainly a reasonable desire.

Air can be warmed from some powered equipment by the engine's hot cooling fluid. The hot coolant is contained in a heater core which looks and acts very much like the engine's radiator. Heat is allowed to leave the engine coolant to move into the cooler surrounding air; resulting in warm air and cooler fluid.

Heating is usually accomplished by using the same heating medium (engine coolant) repeatedly and circulation is possible using the engine's water pump. The basic principles, that a heat engine produces heat and that heat moves from a hot object to another that is cooler, can be used to move heat from the engine to cooling fluid, then from the coolant to the air in the cab. Movement is always from hot to cooler.

Cooling the air is not as easy, because we don't have a cold object or substance readily available to absorb large amounts of heat. A block of ice has been used, but it is possible to provide a cold object in the cab that is much better suited to absorb heat.

Fig. 1-1—Heat moves from hot to cool.

SERVICE · Paragraphs 2-4

THEORY OF OPERATION

HEAT MOVEMENT

2. **Heat always moves from a warmer substance to a colder substance** when two substances of different temperature are close together. Warm water with ice cubes is an example. Heat from the water flows into the ice which has less heat until the ice and the water become the same temperature. The British Thermal Unit (BTU) is defined as a standard of measurement for determining the amount of heat which transfers from one substance to another. One BTU is the amount of heat necessary to raise the temperature of one pound of water one degree Fahrenheit. An example would be 180 BTU's would be required to raise the temperature of one pound of water from 32 degrees F. to 212 degrees F. Conversely 180 BTU's must be removed from one pound of water to lower the temperature from 212 degrees F. to 32 degrees F.

Fig. 1-2—When fluid boils, it absorbs heat without raising the temperature.

LATENT HEAT

3. All substances exist in one of the states of matter: A solid, a liquid or a gas. Water absorbs heat at 32 degrees F. as it changes from a solid (ice) to a liquid and also at 212 degrees F. when it changes from a liquid to a gas (steam). **When a fluid boils (changes from liquid to a gas) it absorbs heat without raising the temperature of the gas. When the gas condenses (changes back to liquid) it radiates heat without lowering the temperature of the resulting liquid.** An example of this would be heating one pound of water in an open container over a flame. The water will raise in temperature one degree F. for each BTU of heat absorbed until the temperature reaches 212 degrees F. The temperature of the water will not increase above 212 degrees F. even if the flame continues to release its heat to the water no matter how hot the flame. The water will boil (change from liquid to gas) and continue to boil until it is all vaporized and leaves the pan. It is possible to collect this vapor in a container and the temperature will still be 212 degrees F. One pound of water at 212 degrees F. will absorb 970 BTU's of heat in changing to a vapor which is also 212 degrees F. and release 970 BTU's when condensing back to a liquid which is also 212 degrees F. The hidden 970 BTU's of heat that can't be measured with a thermometer when boiling the one pound of water is called the "Latent Heat of Vaporization". The movement of this latent heat of vaporization which occurs when any fluid changes from a liquid to a vapor, then back to a liquid is the basis of all conventional refrigeration systems.

For use in a refrigeration system the amount of heat absorbed when vaporizing is important, but also the temperature at which vaporization occurs must be lower than the desired temperature of the substance to be cooled.

Several different fluids have been used, but Refrigerant-12 is exclusively used in farm equipment air conditioning systems. Refrigerant-12, boils at -21.7 degrees F. (-29.8 degrees C), absorbs large amount of heat without changing temperature and when boiling is much cooler than the desired temperature of the air in a farm equipment cab.

EFFECTS OF PRESSURE

4. **The temperature at which a liquid vaporizes or that a gas condenses depends upon the pressure.** Water in an open container vaporizes at 212 degrees F., but in a closed container water can be heated about 2.5-3 degrees F. hotter with each pound per square inch increase above atmospheric pressure. This principle is used in engine cooling systems by installing a pressurizing radiator cap. The water inside the engine's cooling system can be heated well above 212 degrees F. by using a radiator cap which causes 12-14 psi pressure inside the cooling system. Lowering the pressure below atmospheric will cause water to boil at less than 212 degrees F.

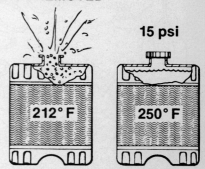

Fig. 1-3—Pressurizing the engine cooling system raises the temperature at which the coolant such as water will boil.

Paragraph 5

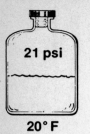
21 psi
20°F

46.6 psi
50°F

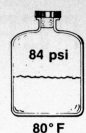
84 psi
80°F

Fig. 1-4 — Raising or lowering the pressure under controlled conditions can be used to change the fluid from liquid to gas, then back to liquid.

Raising and lowering the pressure of refrigerant (R-12) under very controlled conditions can be used to change the state of the fluid from liquid to gas then back to liquid.

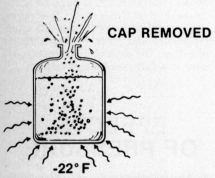

CAP REMOVED
-22°F

Fig. 1-5 — Refrigerant-12 boils at approximately -22 degrees F. when released to atmospheric pressure.

When liquid refrigerant is released into a warm, lower pressure, the temperature of the liquid refrigerant is lowered to -21.7 degrees F. (-29.83 degrees C.) which is its boiling point. It stays at -21.7 degrees F. and boils as long as there is some liquid remaining just as the water in an open pan remains at 212 degrees F. while boiling. This controlled release of liquid R-12 takes place in the air conditioner evaporator. Warm air from inside the cab passes around the evaporator coils and heat moves from the warm air to the boiling liquid, -21.7 degrees F., R-12 inside the evaporator.

RECYCLING REFRIGERANT

5. To reuse the vaporized refrigerant, heat must be removed from the R-12, it must be changed back to a liquid state and must be returned to the point at which it is released to the lower pressure of the evaporator.

A mechanical compressor is used to draw the vaporized (evaporated) refrigerant from the evaporator and force it under high pressure into a condenser. The refrigerant enters the condenser as a high pressure vapor. In-

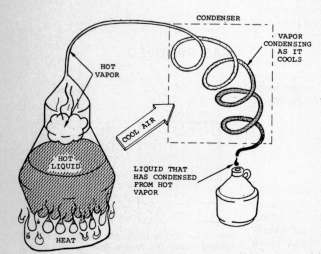

Fig. 1-6 — A condenser is used to contain vapor (gas) until it it cools back to its liquid state.

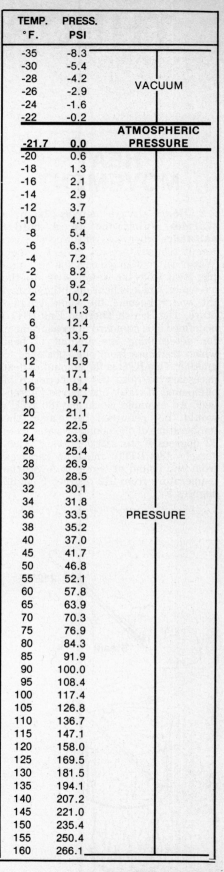

TEMP. °F.	PRESS. PSI	
-35	-8.3	
-30	-5.4	
-28	-4.2	VACUUM
-26	-2.9	
-24	-1.6	
-22	-0.2	
-21.7	0.0	ATMOSPHERIC PRESSURE
-20	0.6	
-18	1.3	
-16	2.1	
-14	2.9	
-12	3.7	
-10	4.5	
-8	5.4	
-6	6.3	
-4	7.2	
-2	8.2	
0	9.2	
2	10.2	
4	11.3	
6	12.4	
8	13.5	
10	14.7	
12	15.9	
14	17.1	
16	18.4	
18	19.7	
20	21.1	
22	22.5	
24	23.9	
26	25.4	
28	26.9	
30	28.5	
32	30.1	
34	31.8	
36	33.5	PRESSURE
38	35.2	
40	37.0	
45	41.7	
50	46.8	
55	52.1	
60	57.8	
65	63.9	
70	70.3	
75	76.9	
80	84.3	
85	91.9	
90	100.0	
95	108.4	
100	117.4	
105	126.8	
110	136.7	
115	147.1	
120	158.0	
125	169.5	
130	181.5	
135	194.1	
140	207.2	
145	221.0	
150	235.4	
155	250.4	
160	266.1	

Fig. 1-7 — The table shows the relationship of pressure and temperature of R-12.

creasing the pressure of refrigerant inside the condenser raises the boiling temperature above the temperature of the outside air. Heat from the refrigerant inside the condenser as always moves to a cooler substance which is in this case outside air. When enough heat is removed from the refrigerant, the R-12 changes back to a liquid (condenses). The liquid now under high pressure moves to the reservoir and filter which is called the receiver-drier. Liquid high pressure refrigerant is permitted to leave the receiver-drier and flow through the expansion valve. The expansion valve controls the amount of R-12 entering the low pressure of the evaporator which repeats the cooling cycle.

Heat is moved from the cab to the outside air by controlling the pressure of the refrigerant in the evaporator and condenser. An expansion valve maintains a low pressure inside the evaporator, lowering the boiling temperature and permitting heat from the relative warm air inside the cab to move into the refrigerant. The compressor raises the pressure refrigerant in the condenser, raising the boiling temperature and permitting heat from the refrigerant to be released into the relatively cool outside air. The expansion valve and compressor mechanically cause pressure changes that permit heat to be absorbed in one and released in another.

EXPLANATION OF BASIC PARTS

6. The four main mechanical components used on automotive air conditioning systems are: Compressor, Condenser, Orifice and Evaporator. Tubes and hoses provide a path through which refrigerant can pass from one component to another. Other components are installed to control or protect the systems.

Air conditioning systems are heat transfer units which operate by moving heat from one location to another. Each component must be carefully selected to maintain the delicate balance necessary for compressing and cooling the refrigerant, moving the refrigerant, lowering the pressure and absorbing heat, then repeating the cycle of events.

COMPRESSOR

7. **The refrigerant is circulated and compressed by a pump called a compressor.** The compressor intakes relatively low pressure (7-50 psi) vaporized refrigerant, then mechanically moves this refrigerant into the condenser. The compressor is able to move significantly more refrigerant than is permitted to pass through an orifice in the system after the condenser. The compressor is therefore able to increase pressure of refrigerant located between the compressor and the orifice to a high pressure of about 180-230 psi.

The intake (low pressure or suction) line to the compressor is the larger of the two lines and should feel cool to the touch when the system is operating. The compressor compacts the heat laden refrigerant into the condenser so the smaller (outlet) line from the compressor will feel warm (HOT) when system is operating. Increasing the pressure also raises the boiling point of the refrigerant. In addition to moving the refrigerant between the various components, the higher pressure concentrates the heat so that the heat can be easily radiated to the cooler outside air. The transfer of heat from the refrigerant to the outside air is accomplished by the condenser.

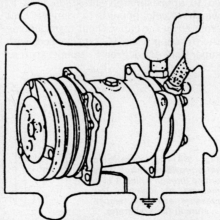

Fig. 2-1 — A compressor is used to move the refrigerant and change the pressure of the refrigerant.

Most air conditioner compressors are two, four, five or six cylinder, reciprocating piston type although other types of compressors have been used. The unit is usually mounted on the side of the engine and driven by the engine crankshaft via one or more "V" belts. Reed valves control the entrance and exit of refrigerant through the compressor. Failure of either intake or exhaust reeds will result in low pressure and improper operation. The compressor, regardless of type, is designed to move compressibles. A plugged orifice prevents movement, overloads the compressor and results in extensive damage. Gaseous vapors can be compressed, but liquid refrigerant cannot. Adding too much refrigerant or any other fault which permits liquid to enter the compressor can easily damage parts.

The compressor is often the only component that needs lubrication. Some compressors are designed with a lubricating sump which should contain sufficient oil to adequately lubricate the compressor. All systems circulate some lubricant throughout the system with the refrigerant; however, this circulating lubricant is the only oil used to lubricate some compressors. The refrigerant used to charge the system is dry (contains no oil) and operating the compressor without lubrication will damage the unit. Be sure that adequate lubrication is present within the complete system before operating the compressor.

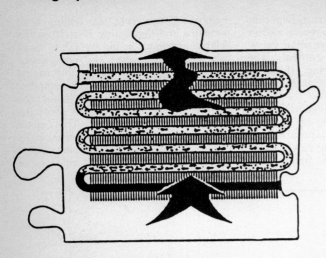

Fig. 2-2 — The refrigerant is allowed to cool (radiate the latent heat) and condense back to a liquid inside the condenser.

CONDENSER

8. The purpose of the condenser is to radiate enough heat from the refrigerant to the air surrounding the condenser to cause the pressurized refrigerant inside to condense (change from vapor to liquid).

Raising the pressure of the refrigerant will raise the boiling point of R-12 just as pressurizing an engine's cooling system by installing a pressure radiator cap will raise the boiling point of water. The compressor pressurizes the vaporized refrigerant and moves it into the condenser where latent heat, stored in the refrigerant, is radiated to the outside air which surrounds the condenser. The refrigerant will change to a liquid (condense) when the temperature is reduced, below the boiling point of the fluid. For instance, the boiling point of R-12 is 134 degrees F. at approximately 190 psi, so the refrigerant will condense when cooled to a temperature of less than 134 degrees F. Heat always moves from the hot substance to the cooler; therefore, heat will leave the hot refrigerant in the condenser and enter the cooler air passing through the condenser. Temperatures and pressures will vary, but the refrigerant must be cooled sufficiently to condense and form liquid before leaving the condenser in order for the air conditioner to cool properly.

ORIFICE

9. A small passage is used to reduce the pressure of the liquid refrigerant as it enters the evaporator. The orifice may be a simple small opening of a specific fixed size of the orifice may be incorporated as part of a control valve known as an expansion valve.

10. An expansion valve controls the opening of the orifice to assure that nearly all of the liquid refrigerant which enters the evaporator is vaporized when it exits the evaporator. The warm liquid refrigerant enters the expansion valve at high pressure and only a small amount of liquid R-12 is allowed to pass through the restriction into the low pressure section. The low pressure will lower the boiling temperature of the refrigerant and heat will be absorbed by the liquid R-12 changing it into a vapor. The boiling point of R-12 at 0 psi is approximately -22 degrees F., but the pressure in the low pressure section of an operating unit will normally be maintained at approximately 15-30 psi. The boiling point of R-12 at 15-30 psi is raised to approximately 11-32 degrees F. which is cold enough to remove heat from air in the cab without experiencing the frosting that is experienced with super cold -22 degree F. temperatures.

Heat is absorbed by boiling liquid refrigerant in the evaporator coil, thus changing the refrigerant to vapor. For maximum evaporator efficiency, some liquid should be supplied throughout the full length of the coils and the pressure inside the evaporator must remain low. The expansion valve or other controls are used to vary the amount of the refrigerant passing through the orifice controls which change the size of the orifice depending upon pressure and temperature in the evaporator. A large volume of liquid R-12 can enter the evaporator when the air around the evaporator (cab) is hot; but a smaller amount of liquid R-12 is needed when the air is cool.

The temperature bulb attached to the expansion valve detects the temperature of the evaporator outlet pipe and varies the opening of the expansion valve accordingly. Gas inside the temperature bulb will expand, if warm, and will open the expansion valve allowing more refrigerant inside the evaporator. Low pressure of refrigerant inside the evaporator and pre-set pressure from a spring are used together to move the valve toward closed. The closing pressure will restrict the size of the valve opening, if the pressure in the evaporator increases.

There are two basic types of expansion valves commonly used. One is internally equalized, the other is externally equalized. The internally equalized type has an internal passage which routes the low pressure from the outlet side of diaphragm of the expansion valve to the underside of control, counteracting the pressure of a warm temperature bulb. The fixed spring pushes the valve closed and is adjusted at the time of manufacture. This spring (called the superheat spring) is temperature sensitive and will prevent flooding of the evaporator by expanding to close the valve.

The operation of the externally equalized expansion valve is like the internally equalized type, except pressure is sensed through a tube attached to the outlet from evaporator.

Fig. 2-3 — The orifice may be a small opening of fixed size as shown or may be part of the expansion valve.

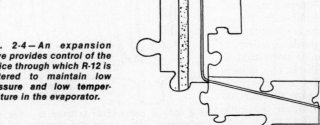

Fig. 2-4 — An expansion valve provides control of the orifice through which R-12 is metered to maintain low pressure and low temperature in the evaporator.

SERVICE

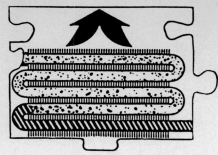

Fig. 2-5 — Refrigerant is contained in the evaporator at low pressure permitting the liquid R-12 to boil (absorb heat) and vaporize or evaporate.

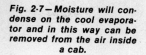

Fig. 2-7 — Moisture will condense on the cool evaporator and in this way can be removed from the air inside a cab.

11. If a fixed orifice is used in place of an expansion valve, pressure and temperature sensitive controls monitor the system, evaluate the condition, then engage or disengage the compressor drive clutch as required to maintain correct and safe control of refrigerant in the evaporator. This is called a Cycling Clutch Orifice Tube (CCOT) type.

EVAPORATOR

12. **Liquid refrigerant which passes the orifice enters the evaporator where the low pressure refrigerant absorbs heat.** The refrigerant inside the evaporator is changed just the opposite of changes in the condenser. The lowered pressure liquid refrigerant in the evaporator absorbs enough heat from the surrounding air to boil the liquid into vapor. Heat always moves from warmer substances to cooler. Heat moves from surrounding air to the metal of the evaporator, then from the evaporator to the low pressure refrigerant inside. Fins are attached to the evaporator core to increase the surface area of the evaporator and blower fans are used to circulate more of the warm air to the evaporator surfaces.

Air is also dehumidified by condensing moisture on the cool outside surfaces of the evaporator in much the same way as

Fig. 2-8 — The speed of evaporation depends upon the temperature difference.

moisture condensing on the sides of a glass of ice water on a warm day. Drains are provided in all air conditioner systems to funnel the condensed moisture out of the cab.

Evaporation of refrigerant inside the evaporator and condensation of refrigerant inside the condensor both depend upon wide differences in temperature to affect the desired exchange of heat. Slowing the movement of cool air over the condenser will slow the radiation of heat, but slowing the movement of air over the evaporator will allow more time for the warm air to give up its heat to the refrigerant. Slower evaporator fan speeds will allow greatest heat absorption resulting in the coolest temperature from the air leaving the evaporator. High blower speeds can be successfully used to circulate extremely hot air over the evaporator core fins such as encountered when the system is first turned on, but continued operation at high blower speeds will not cool the air effectively.

The temperature and boiling point of a liquid raise and lower together. At 0 psi, the boiling point of liquid R-12 is -22 degrees F., but at 30 psi the boiling point is raised to 32 degrees F. Moisture that collects on the evaporator fins will freeze if permitted to cool below 32 degrees F. and the frozen moisture would block air flow which would result in not cooling the cab air. By permitting the pressure in the evaporator to remain at about 15-30 psi, the boiling point of the R-12 is raised to 11-32 degrees F. The moisture that forms is permitted to drain quickly from the outside of the evaporator before it becomes cold enough to freeze at the just below freezing temperatures. The 11-32 degree F. boiling point is still well below the temperature desired.

REFRIGERANT

13. **The fluid used to move heat in automotive air conditioning systems is dichlorodiflurometane (CCl_2F_2) commonly called R-12 or refrigerant-12.** The name "Freon" and "Freon-12" are registered trademarks and should only be used referring to this type of refrigerant produced by DuPont. Other fluids have been used in different refrigeration applications, but **DO NOT**

Fig. 2-6 — On a warm humid day, moisture will condense on the outside of a sufficiently cool object.

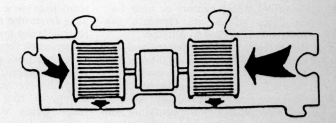

Fig. 2-9 — Blower fans are used to move air around the evaporator and condenser to change the state of the refrigerant.

AIR CONDITIONING

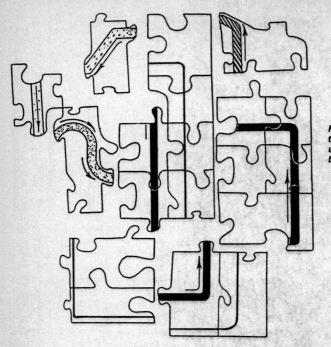

Fig. 2-10 — Various sizes and types of hoses and tubes are used to route the refrigerant between the air conditioner components.

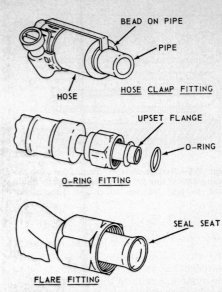

Fig. 2-11 — View showing three types of fittings commonly used in air conditioning systems. Use only an approved refrigerant oil when assembling.

substitute any other refrigerant in a farm equipment air conditioner system. Refrigerant-12 has been accepted because of the high safety factor together with the capability to withstand high pressure and temperature without decomposing or deteriorating. The refrigerant does not react to metals such as aluminum, copper, iron or steel unless exposed to moisture inside the sealed system. An acid is formed when R-12 absorbs moisture in an action called HYDROLYZING. Since air contains moisture, it is important to remove all air (and moisture) from the system as well as water, wet leaves, etc.

Refrigerant-12 is soluable in oil which facilitates lubrication of the compressor. Rubber hoses can be used as flexible connections because rubber does not harm R-12 and the refrigerant does not damage the rubber. Refrigerant-12 is non-flammable as either a liquid or a gas, but exposure of R-12 to an open flame or very high temperature will form a toxic (PHOSGENE) gas which is very dangerous. The refrigerant even as a vapor is heavier than air and extreme caution should be exercised when releasing to the atmosphere to prevent exposure to open flames near the floor such as hot water heaters or near furnace ducts which can circulate the vaporized refrigerant to a flame.

Refrigerant-12 is transparent and colorless and under normal conditions it is considered safe, stable and non-explosive, but control must be maintained to insure safety and stability. Heating the refrigerant in a closed container will increase pressure sufficiently to burst the container. Uncontrolled release of the refrigerant is dangerous because of the extremely cold temperature and possible exposure of the gas to a flame, but the exploding fragments of the container can also cause serious injury. The heat from exposure to direct sunlight (such as in a window display) or from hot exhaust systems can increase pressure enough to burst container. Always store refrigerant in a cool, dry place. The pressure of approximately 200 psi in the high pressure section of an operating air conditioning system may be enough to burst the refrigerant storage container. Be especially careful to **NEVER** connect the refrigerant container to the high pressure section when air conditioner is operating. Always use care when handling the refrigerant.

HOSES, TUBES AND CONNECTIONS

14. **The hoses, lines and connections provide a path for refrigerant to pass from one component to another.** Rubber hoses are used where movement necessitates flexibility; steel, aluminum or copper lines may be used where rigidity is required. The attachment may be any of several types, including hose clamp, "O" ring and flange or flare fitting. Quick disconnect, self-sealing couplings may be used at some points to facilitate service.

The line from evaporator to the compressor inlet carries low pressure vapor. This line is subjected to 7-30 psi during normal operation, but is commonly called the suction line. The suction line should be cool when the system is operating properly.

The line from the compressor outlet to the inlet of condenser contains high pressure compressed refrigerant vapor. Normal pressure in this line is 185-205 psi when the outside air temperature is 95 degrees F., but normal pressure will increase to 250-270 psi when outside air temperature is 110 degrees F. This line is called the discharge line or high pressure vapor line and it will be hot when the system is operating properly.

The lines from the condenser to the receiver-drier and from the receiver-drier to the orifice (or expansion valve) contain high pressure liquid refrigerant. These lines are often called liquid or high pressure liquid lines and both should normally be warm to the touch. Pressure will be just about the same as in the discharge line in a normal operating system.

The line from the orifice to the evaporator is often an extension of the evaporator tubing and should be the coolest part of the system. Pressure is reduced in the expansion valve so the refrigerant is approximately 7-30 psi.

Restrictions of any kind can often be identified by the change in temperature of the line and will probably effect cooling.

SERVICE

Paragraphs 15-16

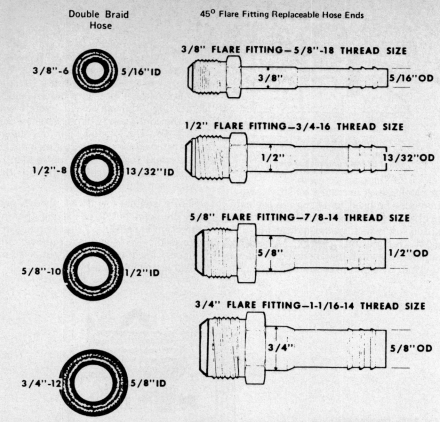

Fig. 2-12 — Refrigeration hose is usually one of the four more popular sizes. Replaceable hose ends must have correct size of hose installed with correct clamp.

ped by passing the refrigerant by a desiccant. The desiccant is sometimes a bag of silica-gel or similar material sealed inside the receiver or accumulator during manufacture, but regardless of the type used, the desiccant will absorb only a small volume of moisture. Water mixed with refrigerant even in very small amounts cannot only damage metal parts by hydrolysis, but can freeze if permitted to flow into low temperature sections of the air conditioner. Ice is of course hard and will block flow of refrigerant as can any solid object within the system.

It is recommended that a new receiver-drier (or accumulator-drier) be installed every time the system is opened for service in addition to complete evacuation (drying) to insure that no moisture be contained in the system.

RECEIVER OR ACCUMULATOR

15. A receiver is often incorporated between the condenser and the expansion valve orifice to assure that the orifice is supplied with cool liquid refrigerant. Conversely an accumulator is often located between the outlet from the evaporator and inlet to the compressor to be sure that all refrigerant entering the compressor is vaporized. The quantity of refrigerant stored in the receiver or accumulator during normal operation will compensate for changing demands caused by varying heat load, condensing action and compressor speed. The additional refrigerant will also permit loss of small amounts of refrigerant without impairing cooling.

It is important to notice the design differences. The accumulator, located between the evaporator and the compressor inlet, should only pass vaporized refrigerant to prevent damage to the compressor. The receiver, located between the condenser and the orifice or expansion valve, should pass liquid refrigerant. It is also important that units be installed correctly so that features will work correctly. Mounting the receiver or accumulator at an improper angle can prevent correct operation.

Often screens or filters and driers are incorporated into the design of receivers and accumulators. Moisture can be trap-

FILTERS

16. **It is important to remove even very small contaminants from the refrigerant.** Filter screens are located in several locations within the system. Filters are located in the receiver or accumulator, at the orifice or expansion valve and at the inlet port of the compressor. Some filter screens may be removed and cleaned, but the type sealed inside the sealed receiver or accumulator can be serviced by installing a new unit.

Restrictions of any kind can often be identified by the change in temperature of the line and will probably effect cooling.

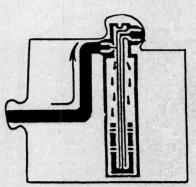

Fig. 2-13 — A receiver may be included in the high pressure liquid part of the system to provide an extra reserve of liquid refrigerant in normal operation.

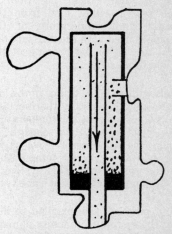

Fig. 2-14 — An accumulator may be included in the low pressure vapor part of the system to assure that liquid refrigerant does not enter the compressor.

Paragraph 17

AIR CONDITIONING

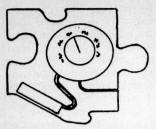

Fig. 2-16 – Controls may sense the temperature of various components or the air temperature, then adjust operation of the cooling system.

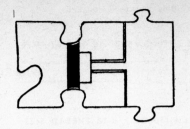

Fig. 2-17 – Controls which sense the pressure of refrigerant may be used to control the engagement of the compressor clutch.

CONTROLS

17. A variety of different controls are necessary to provide desired air temperature control as well as proper temperature and pressure of the system refrigerant.

The temperature and boiling point of a liquid raise and lower together. At 0 psi, the boiling point of liquid R-12 is -22 degrees F., but at 30 psi the boiling point is raised to 32 degrees F. Moisture that collects on the evaporator fins will freeze if permitted to cool below 32 degrees F. and the frozen moisture would block air flow which would result in not cooling the cab air. By permitting the pressure in the evaporator to remain at about 15-30 psi, the boiling point of the R-12 is raised to 11-32 degrees F. The moisture that forms is permitted to drain quickly from the outside of the evaporator before it becomes cold enough to freeze at the just below freezing temperatures. The 11-32 degree F. boiling point is still well below the temperature desired in the cab.

On models equipped with an expansion valve, a temperature sensing bulb and capillary tube are used to control the opening of the orifice. The bulb is clamped to the outlet pipe from the evaporator and temperature of the outlet pipe is used with the pressure differential to control the opening of the expansion tube orifice.

Models with a simple orifice in the line to the evaporator use a thermostatic control switch to sense the temperature of the evaporator, then engage or disengage the compressor drive clutch to maintain the desired temperature within the evaporator. This system is called Cycling Clutch Orifice Tube type.

Safety switches are installed to **help** protect the system from major damage. A pressure sensitive control is used to disengage the compressor clutch if the refrigerant high pressure is too high which may be caused by excessive heat or blockage. A pressure sensitive switch may also be used to disengage the compressor drive clutch if the high pressure is too low which would indicate insufficient refrigerant charge. A special superheat switch may be used to blow a thermal fuse and prevent damage to the clutch and compressor in case of low refrigerant charge. A typical superheat switch as shown in Fig. 3-10 is installed in the inlet passage of the compressor and senses both temperature and pressure of the refrigerant. If the temperature and pressure become sufficiently low, the superheat switch will close the contact to ground. As shown in Fig. 3-10A, the ground from the superheat switch causes the heater element of the thermal fuse to melt, permanently stopping current from engaging the compressor clutch.

Safety switches will not always prevent damage, but it is certain that no protection is possible if the switches are by-passed. Be sure to remove any jumpers, used to troubleshoot system, before returning to service.

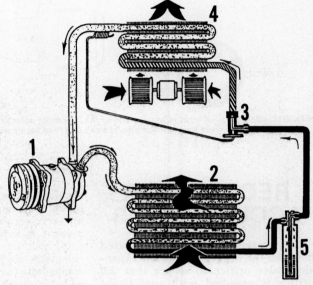

Fig. 2-18 – Drawing of a complete air conditioning system showing direction of refrigerant flow. The system shown uses an expansion valve as shown at (3).

1. Compressor
2. Condenser
3. Expansion valve
4. Evaporator
5. Receiver

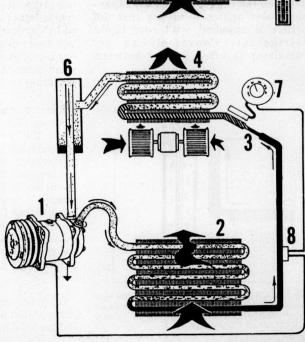

Fig. 2-19 – Drawing of air conditioning system typical of types using a fixed orifice (3). Temperature at evaporator inlet is sensed by thermostatic control (7) and pressure of high pressure line is sensed by control (8).

1. Compressor
2. Condenser
3. Orifice
4. Evaporator
6. Accumulator
7. Thermostatic switch
8. Pressure switch

SERVICE

Paragraphs 18-19

UNIT APPLICATION

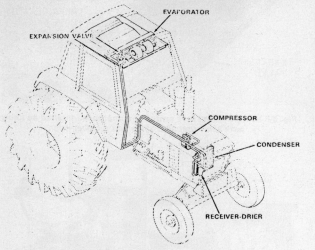

Fig. 3-1 — View showing the layout of a typical air conditioner installed with a tractor cab.

18. The location and the type of components used will depend upon the specific requirements of the tractor, combine or other equipment upon which the air conditioner is installed as well as the manufacturer of the cab and/or air conditioner system.

The compressor is driven from the engine by a "V" belt so the compressor must be located near the accessory drive end of the engine. The evaporator removes heat from air in the cab so it will be in or near the passenger compartment. The expansion valve will be located at the inlet to the evaporator. The condenser can be located about anywhere there is a flow of cooling air necessary for radiating heat from the refrigerant. Sometimes the condenser is located ahead of the engine's cooling system radiator so that the engine fan will circulate air through the condenser as well as the radiator. The condenser can also be located on the roof or some other position and electric fans can be used to circulate air through the condensor coils. The receiver is used to maintain an even flow of liquid refrigerant to the orifice and is therefore located in the liquid high pressure line between the condenser and the orifice. Usually the receiver will be located near the condenser outlet. If so equipped, the accumulator is used to assure that only vaporized refrigerant enters the compressor. Any liquid refrigerant is accumulated in the unit where it is permitted to vaporize before entering the compressor. The accumulator is located in the low pressure vapor line between the evaporator outlet and the inlet port of the compressor, usually near the evaporator outlet. If equipped with a sight glass, the sight glass may be made as part of the receiver or may be located in the high pressure liquid line between the condenser and the orifice.

Fig. 3-2 — View of typical roof mount air conditioner system. Compressor, which is mounted on engine, is not shown.

1. Motor & squirrel cage blowers for evaporator
2. Motors and fans for condenser
3. Air filter
4. Evaporator
5. Condenser
6. Receiver-drier unit
7. Expansion valve
8. High pressure safety switch
9. Low pressure safety switch

CONTROLS

19. Operator control of the air conditioning system is usually limited to the setting of two or three switches and a temperature control thermostat all located within the cab. Additional automatic controls are used to reduce damage and increase safety in case of system failure.

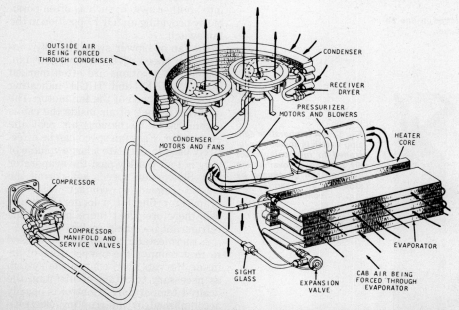

Fig. 3-3 — Drawing showing components of a typical air conditioner system.

Paragraph 19 Cont.

AIR CONDITIONING

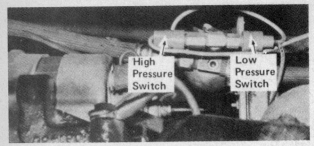

Fig. 3-4 — View of high pressure and low pressure safety switches installed.

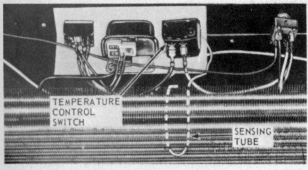

Fig. 3-5 — View of temperature control switch and its sensing tube installed.

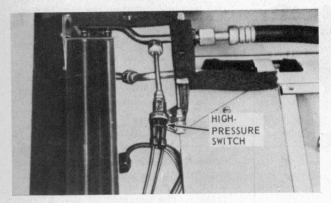

Fig. 3-6 — High pressure switch is sometimes installed in line between compressor discharge and condenser.

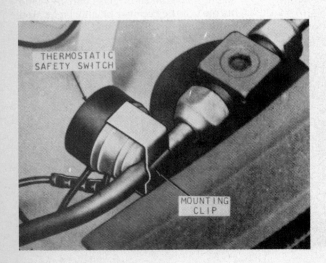

Fig. 3-7 — A temperature sensing switch is sometimes installed to turn compressor off in case of overheating such as can be caused by stopping air flow through condenser.

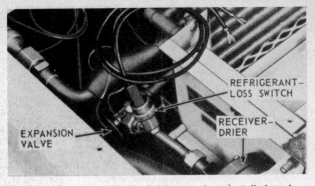

Fig. 3-8 — A refrigerant loss switch is sometimes installed as shown between expansion valve and receiver-drier.

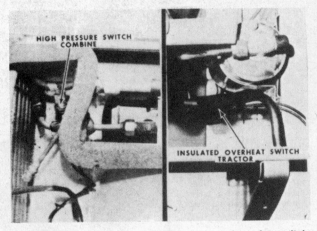

Fig. 3-9 — View of pressure and temperature sensing safety switches installed.

The main control switch intercepts electrical power to all parts of the air conditioner when in the "OFF" position. This main switch may be incorporated into another switch, such as often possible by providing an "OFF" position in the fan switch.

The fan or blower switch usually controls the speed of the evaporator fan. The switch positions are often marked "LOW", "MED." and "HIGH" indicating the general speed of the fan motor. The effective cooling of air inside the cab is accomplished by moving air past the evaporator. Fast fan speed can be effectively used to circulate a large volume of very hot air quickly past the evaporator fins. Slower fan speeds will permit the air to remain in contact with the evaporator fins for a longer time and will therefore yield more heat to the refrigerant becoming cooler.

A thermostatic control switch is used to first monitor the temperature of air inside the cab, then control the system as necessary to attain or maintain the desired temperature. Control is usually accomplished by interrupting electrical power to the compressor drive clutch.

SERVICE

Paragraph 19 Cont.

Stopping the compressor by disengaging the drive clutch will cause liquid refrigerant to stop flowing into the evaporator and therefore cooling will cease.

Another method of maintaining a desired temperature is by mixing heated air with the cooled air. In automotive application, control of the necessary ducting is accomplished by vacuum operated mechanisms. On systems which are not designed to mix heated and cooled air, accidentally leaving the heater on can be confused for inadequate cooling.

Fuses, circuit breakers and automatic switches may also be included in the

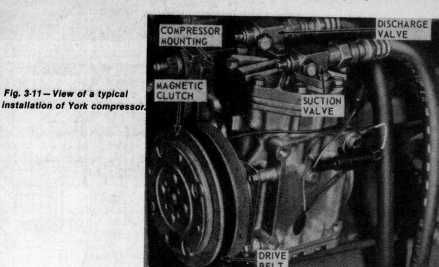

Fig. 3-11 — View of a typical installation of York compressor.

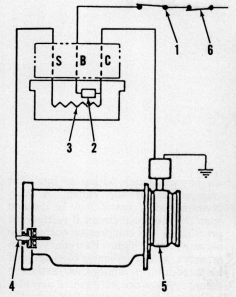

Fig. 3-12 — View of two installations of Tecumseh compressors.

Fig. 3-10 — Drawing showing installation of superheat switch (4), thermal fuse (2) and associated parts. High pressure (6), low pressure (1), and various temperature switches may be installed in the circuit to disengage the compressor drive clutch.

1. Low pressure switch
2. Thermal fuse
3. Heater
4. Superheat switch
5. Clutch
6. High pressure switch

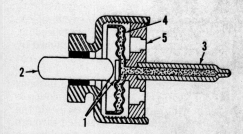

Fig. 3-10A — Cross-section of superheat switch shown at (4) in Fig. 3-10. Both high temperature and low pressure are necessary to force contacts (1) together.

1. Contacts
2. Terminal
3. Sensing tube
4. Diaphragm
5. Openings

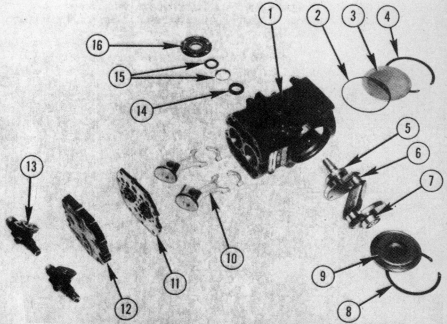

Fig. 3-13 — View of disassembled Tecumseh compressor.

1. Crankcase
2. "O" ring
3. Bottom plate
4. Snap ring
5. Front bearing
6. Crankshaft
7. Rear bearing
8. Snap ring
9. Rear bearing housing
10. Connecting rod & piston
11. Reed plate
12. Cylinder head
13. Service valve
14. Seal gland
15. Carbon ring
16. Seal plate

15

AIR CONDITIONING

Paragraph 19 Cont.

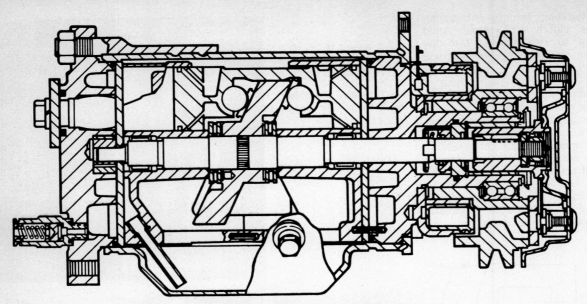

Fig. 3-14—Cross-section drawing of Delco Air compressor.

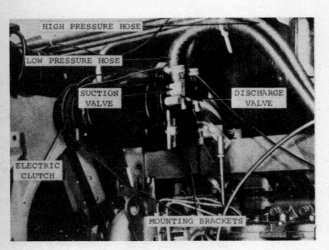

Fig. 3-15—View of Delco Air compressor installed.

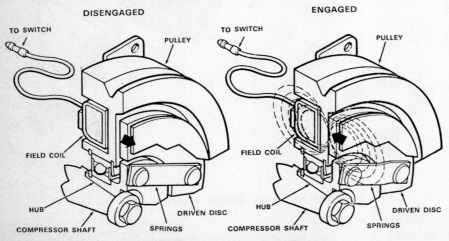

Fig. 3-16—Drawing showing principle of magnetic clutch operation.

compressor clutch wiring to interrupt engagement of the clutch. A high pressure safety switch may be used to open the electrical circuit if refrigerant pressure in the compressor output line becomes too high. Extremely high pressure usually indicates blockage and may result in injury or extensive damage. A low pressure switch may also be located in the output line from the compressor to prevent engagement of the compressor clutch if pressure is too low. Extremely low pressure would indicate an insufficient charge of refrigerant and operation would severely damage the compressor. A superheat switch may be installed in conjunction with a thermal fuse as shown in Fig. 3-10 to protect against overheating. Special pressure releasing blow out fuses are used on some systems to release refrigerant in the event of excessive pressure in the low pressure line. Excessive pressure is usually caused by overheating of a system which is not operating.

Other controls may include a damper or air control knob which controls whether the air is recirculated inside the cab or if outside air is circulated through the evaporator. A switch is provided on some models to reverse the direction of air flow through the condenser by reversing the condenser fan motors. Switches such as heater, window wiper, etc., may also be located on the control panel.

SERVICE

COMPRESSOR

20. The compressor may be any of several types. The most commonly used compressors for automobiles, trucks and farm equipment air conditioners are: Delco Air (Frigidaire) Model A-6 with six axial pistons; or Model R-4 with four radial pistons; Sankyo (Abacus), with five axial pistons; Tecumseh, with two pistons; York models with two pistons. All units are designed to compress vapor and are damaged by non-compressibles such as dirt, moisture, liquid refrigerant, etc. The compressor draws vaporized refrigerant from the evaporator, which maintains the low pressure necessary for proper evaporation; compresses the vapor to a high pressure, which is necessary for condensation; and moves the high pressure vapor into the condenser, where heat can be radiated for changing the refrigerant back to a liquid.

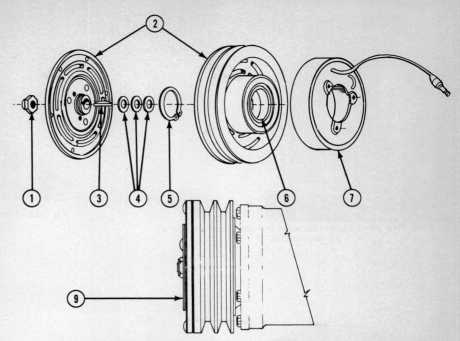

Fig. 3-17 – Exploded drawing of clutch typical of type used with Sankyo compressor.

COMPRESSOR DRIVE CLUTCH

21. Air conditioner systems use an electrically actuated clutch to engage and disengage drive to the compressor. The "V" belt pulley is mounted on a bearing and is free to rotate without turning the compressor crankshaft anytime electrical power is disconnected. The compressor is not operating when the pulley is free wheeling. The field coil is mounted stationary to the front of compressor. The coil is energized by supplying electrical current to one end of wire, the other end of coil winding is grounded to compressor and equipment frame. Energizing the coil creates a magnetic force which locks the driven disc to the pulley and drives the compressor.

The temperature control and safety controls are usually electrical switches that control air conditioner operation by engaging or disengaging the compressor clutch.

Operation of a Cycling Clutch Orifice Tube system is controlled by engaging or disengaging the compressor clutch to maintain the correct system pressure.

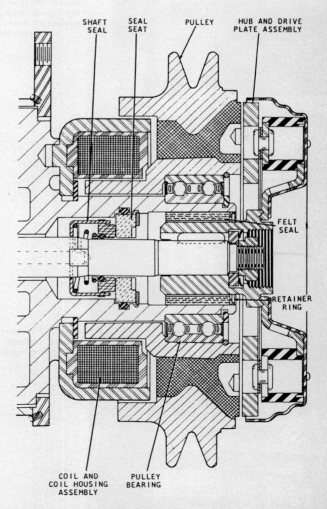

Fig. 3-18 – Cross-sectional of magnetic clutch used with Delco Air compressor.

Paragraph 22   AIR CONDITIONING

Fig. 3-19—View of condenser installed in front of regular cooling system radiator.

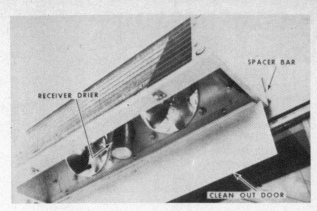

Fig. 3-22—View showing a typical receiver-drier installation.

Fig. 3-20—View of condenser installed on cab. Fan used to circulate air through condenser coils can be seen just inside housing.

Fig. 3-21—View of condenser assembly shown in preceding Figure.

CONDENSER

22. The purpose of the condenser is to radiate enough heat from the vaporized refrigerant to change it to liquid. During normal operation the whole high pressure section will be warm to hot, but large quantities of heat should be radiating from the condenser. Nothing should be permitted to stop or slow down this radiation of heat. Fins are located on the condenser tubes and fans are used to circulate more cool air around the condenser tubes. Keep all leaves, paper, dirt etc., cleaned from condenser and condenser filter screens. The fins on condenser should be straight to permit free flow of air. The condenser is sometimes located ahead of the engine radiator and blockage of air flow through the radiator will also affect the condenser. Be sure to check for leaves, etc., between radiator and condenser. On all models, bent fan blades, slipping fan drive, inoperative condenser fan drive motors on any other fault that lessens the amount of cool air circulated through the condenser should be corrected. The fins and condenser tubes should be clean; because, even a thin layer of oil, dirt or antifreeze will act as an insulator that will slow down the radiation of heat.

Since the purpose of the condenser is to radiate heat, anything that prevents or slows down this action may affect cooling, but the temperature and pressure of the refrigerant raise and lower together. Heat that has not been radiated will remain in the refrigerant and the result will be pressure that is too high. The condenser, hoses, connections and seals can be damaged by the high pressure. Temperature sensing safety switches may be operated by the high temperature caused by the condenser not radiating enough heat.

SERVICE

RECEIVER-DRIER

23. Models equipped with a liquid receiver located between the condenser and the expansion valve, usually incorporate a drier as part of the receiver-drier unit. Moisture is the major enemy of the air conditioning system and the desiccant inside the receiver-drier will absorb only a small amount. Many manufacturers recommend installation of a new receiver-drier each time any part of the refrigeration system is open to the atmosphere. The container of desiccant inside receiver-drier may break open and contaminate the system, if any attempt is made to dry the desiccant or if more moisture is inside system than the desiccant can absorb.

Systems utilizing a Cycling Clutch and Orifice Tube (CCOT) are not usually equipped originally with a receiver-drier filter between the condenser and the orifice tube; however, after market kits are available from usual parts sources for installing one. Installation offers protection from damage caused by moisture or contamination in the orifice tube and evaporator.

On all models, every effort should be made to remove all moisture from the system and install new, dry receiver-drier if condition is questionable. Connections of new receiver-drier should be securely capped before installation to prevent the entrance of moisture (air) while in storage.

A filter screen is located in the receiver-drier to stop solid contaminants from leaving the unit. Blockage of the filter may result in a drop in pressure that will be indicated by a lower temperature at the outlet fitting.

EXPANSION ORIFICE

24. Expansion orifices may either be a restriction with a fixed size of opening or may include controls to open and close the restriction in which case the expansion orifice is called a thermostatic expansion valve.

A fixed orifice is a specially designed restriction which is inserted into the line which enters the evaporator. Control of the refrigerant is maintained by a thermostatic switch which interrupts electrical current to the compressor clutch. The system is referred to as Cycling Clutch and Orifice Tube or C.C.O.T. type.

Systems using thermostatic expansion valves use basically one of two types of valves: Internally equalized and Externally equalized. The two types of expansion valves are similar, but not interchangeable. Both types of expansion valves are installed in the system to lower the pressure before the refrigerant enters the evaporator. The reduction in pressure is accomplished simply by passing the refrigerant through a small hole (orifice), but the opening and closing of the orifice must be controlled to compensate for changes

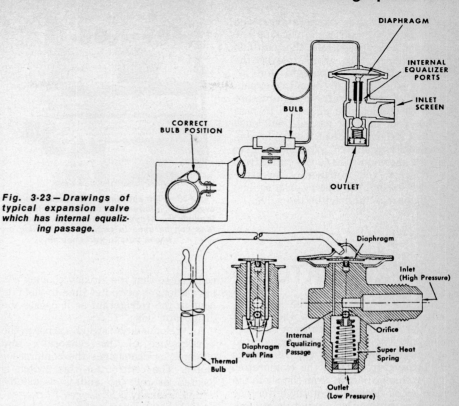

Fig. 3-23 — Drawings of typical expansion valve which has internal equalizing passage.

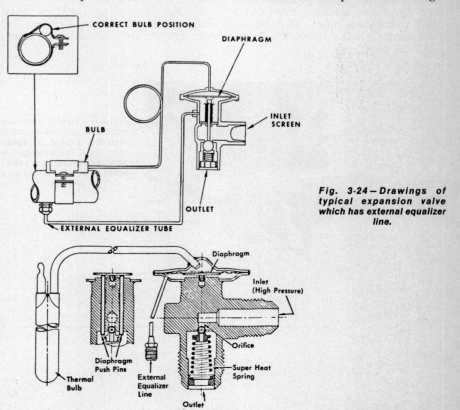

Fig. 3-24 — Drawings of typical expansion valve which has external equalizer line.

in pressure and temperature. The temperature of refrigerant leaving the evaporator is sensed by a thermal bulb and capillary tube which moves the valve seat via a diaphragm and actuating pins. Internally equalized expansion valves permit refrigerant pressure from the outlet side of orifice to pass through an internal passage and push against the underside of the diaphragm. Externally equalized expansion valves have a line connected to the outlet from evaporator and refrigerant pressure passes through this line to push against the diaphragm and act on the valve.

Fig. 3-26 — View showing a typical installation of evaporator. Blower fans circulate air from cab through the evaporator. The same circulating fans can be used to circulate air through the heater core to warm cab air.

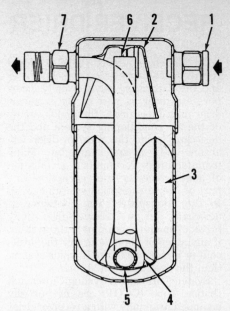

Fig. 3-27 — Cross-section of accumulator showing typical construction. Opening (6) of outlet tube is covered by baffle (2). Small hole (5) maintains lubricant level within the accumulator.

1. Inlet
2. Baffle
3. Desiccant
4. Filter
5. Oil hole
6. Outlet opening
7. Exit connection

EVAPORATOR

25. The evaporator is the low pressure, low temperature area where liquid refrigerant absorbs heat from surrounding air. The refrigerant is evaporated or vaporized by the heat which it absorbs while passing through the evaporator. The exchange of heat from the air to the refrigerant depends upon the difference in temperature. During high heat load, such as usually encountered when the system is first turned on, the temperature difference is great and the refrigerant will absorb heat quickly. The blower fan can be set at its fastest setting to circulate large quantities of warm air around the evaporator under high heat load conditions. After the air has cooled, the fan speed should be reduced so that the already cool air will have a longer period of time to yield its heat to the refrigerant as it passes the evaporator coils.

A thermostatic switch senses the temperature of the evaporator and engages or disengages the compressor clutch. The control for this switch is located in the cab and is adjustable within normally a 12 degree F. range.

ACCUMULATOR-DRIER

26. Models equipped with a fixed orifice (Cycling Clutch and Orifice Tube) type system, are equipped with an accumulator between the outlet from the evaporator and the inlet to the compressor. The major purpose of the accumulator is to make sure that liquid refrigerant does not enter and damage the compressor. The accumulator is also a convenient location for installation of desiccant. Moisture within the system will cause extensive damage and every effort should be made to remove all moisture from the system. The small container of desiccant inside the accumulator will only absorb a small amount of moisture and a new accumulator-drier should be installed before the desiccant has absorbed its limit. Any attempt to dry the desiccant with heat or alcohol will probably contaminate the system. Many manufacturers recommend installation of a new accumulator-drier each time any part of the system is opened to the atmosphere. Be sure that new unit is securely capped before installation to prevent entrance of moisture (air) while in storage.

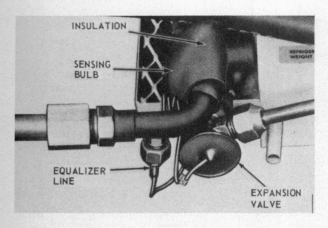

Fig. 3-25 — View of externally equalized expansion valve installed. Except for equalizer line installation is similar for expansion valves with internal equalizing passage.

SAFETY

27. The engine must be running in order to operate the air conditioner during many of the tests and during some servicing. Caution must be exercised around any running equipment especially when its operation is secondary to locating a problem relating to another system. **Don't forget that the engine is running.** Even while working on the air conditioner system, exercise care about engine related safety hazards such as: Exhaust gasses, cooling fan, drive belts, electrical shorts, hot engine cooling and exhaust systems etc.

Refrigerant-12 is easier to handle safely than other fluids that have been used in refrigeration systems, but incorrectly used, R-12 can be dangerous. Be sure to observe the following precautions:

WEAR SAFETY GOGGLES

28. **Avoid any physical contact with the refrigerant.** The liquid refrigerant boils at -22 degrees F. (-30°C.) at which time it is vaporized and the rapid evaporation freezes the area. Contact with any exposed skin can cause frost bite, but serious eye damage or blindness can result from contact with eyes. Use eye wash immediately if accident occurs.

DO NOT OVERHEAT CONTAINER

29. **Do not allow refrigerant to get hot.** Temperatures above 125 degrees F. (52°C.) may increase pressure inside container sufficient to burst the container. Pressure inside container increases rapidly above 125 degrees F. and in a container full of liquid R-12, the pressure may reach 1600 psi before the temperature reaches 165 degrees F. Damage and injury can result from exploding container, refrigerant contacting the skin and/or refrigerant contacting source of heat and changing to poisonous gas.

Don't heat container even when servicing system unless pressure inside can is monitored with a gage. Don't incinerate even empty cans.

Be careful about using heat around an installed air conditioning system for the same reason. To avoid a dangerous explosion, never weld, use a blow torch, steam clean, or use any excessive amount of heat on, or in the immediate area of any part of the air conditioning system.

DO NOT OPEN STORAGE CONTAINER TO THE HIGH PRESSURE SIDE OF SYSTEM WITH AIR CONDITIONER OPERATING

30. The container used to store refrigerant may burst at pressures above 170 psi and normal pressure for the high side of system operating at ambient temperature of 110 degrees F., is 250-270 psi.

DO NOT FILL WITH LIQUID REFRIGERANT ANYTIME COMPRESSOR IS OPERATING

31. **Keep refrigerant containers upright when charging the system.** If the container is on its side or upside down, liquid refrigerant may enter the system and damage the compressor.

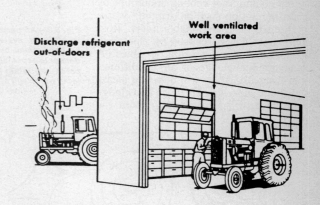

Fig. 4-2 — Release refrigerant only in a safe well ventilated area.

DO NOT BEND OR TWIST ANY OF THE METAL LINES WHILE SERVICING

32. Cracks, broken lines or loose connections can be caused by improper servicing. The resulting loss of refrigerant may cause injury or damage.

RELEASE REFRIGERANT SAFELY

33. The liquid refrigerant expands rapidly when released to the atmosphere and will displace the air. To prevent possible suffocation in enclosed areas, **always discharge the refrigerant slowly and into an approved area** such as into a shop exhaust collector. Always maintain adequate ventilation surrounding work area. The vaporized refrigerant is heavier than air and will collect in low areas.

Do not allow the system to discharge rapidly as this would sweep some of the refrigerant oil out of the compressor. In case of failure which rapidly discharges the refrigerant, check the oil level before recharging.

USE EXTREME CAUTION AROUND OPEN FLAME

34. **The fumes produced by burning refrigerant are a very poisonous (Phosgene) gas** in addition to the normal fire hazard using any open flame. Be especially careful operating leak detectors which use a lighted flame, because refrigerant, explosive gas, dust, etc., is drawn into the flame as a method of checking for leaks.

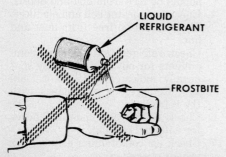

Fig. 4-1 — Exposure to the liquid refrigerant can cause frostbite and is extremely dangerous if improperly used.

TROUBLESHOOTING

35. Problems with an air conditioner system will usually present themselves by not cooling, not cooling enough or by irritating action such as dripping water in the cab or objectionable noise. Most complaints can be caused by electrical, mechanical or refrigerant troubles. A thorough knowledge of air conditioner operating principles will help you locate a problem more quickly. This knowledge together with specific details about the unit will speed troubleshooting and limit the amount of duplication and unnecessary work. Often one problem will result in a chain reaction of failures and the initial cause must be located and corrected to prevent the trouble from recurring. **Be sure to check simple things first.** Do not being installing air conditioner components to correct a complaint about lack of cooling, if the heater is also on.

Fig. 5-3 — All parts of the air conditioner system should be kept clean, but the condenser is especially prone to catch and retain foreign objects which will affect operation.

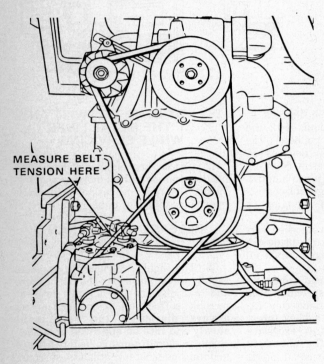

Fig. 5-1 — Compressor drive belts must be tight and in good condition to prevent slippage.

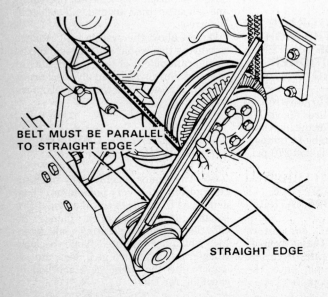

Fig. 5-2 — Be sure that compressor is mounted correctly so that drive pulleys are exactly aligned. Even slight misalignment can quickly destroy drive belts.

PRELIMINARY CHECKS

36. The following checks should be performed before starting the system tests.
1. Check compressor drive belt for proper tension.
2. Check compressor pulley for alignment and compressor mounting bracket for tightness.
3. Check condenser coil fins. They should be free from dirt and debris.
4. Check evaporator coil and air filter. They should be free from dust and lint.
5. Check refrigeration lines and components for oily spots. An oily spot usually indicates refrigerant leakage. Compressor oil escapes with the refrigerant.
6. Check refrigerant lines and hoses for cuts and chafing.
7. Check air conditioning controls for proper operation.
8. Check blower fan operation at all speeds.

SERVICE

Paragraphs 37-39

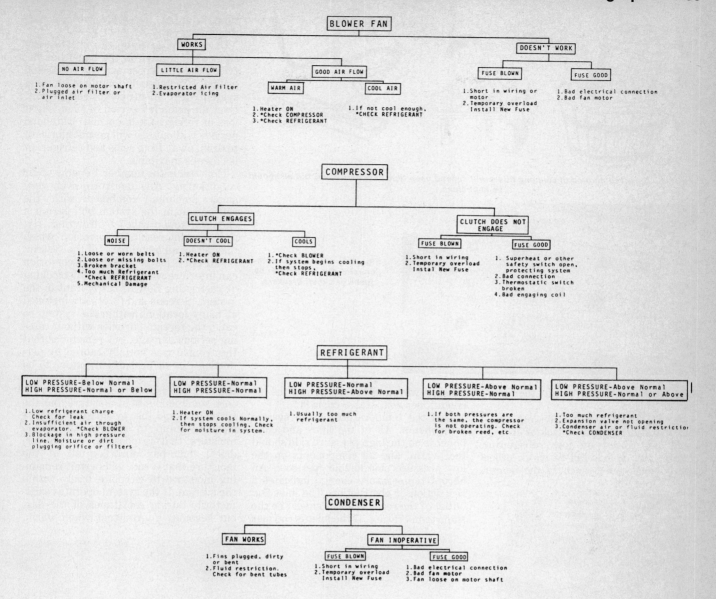

OPERATIONAL CHECK

37. The first check should be to observe the symptoms by starting the engine and attempting to operate the air conditioner normally.

The evaporator blower, the compressor clutch and the condenser fans (if so equipped) should all operate when the air conditioner is turned on. The compressor drive clutch should engage with a snap which can be heard near the unit and the drive plate at the front of the clutch should begin to rotate with the pulley as the clutch is engaged.

38. **LACK OF ACTIVITY.** A complete lack of activity will usually indicate an electrical power supply problem such as a blown fuse, corroded terminal, broken or disconnected wire, bad main switch, etc.

If a fuse is blown, the cause of the blown fuse should be located and corrected. If the system is shorted, installing a new fuse will not solve the problem.

Sometimes electrical problems will be accompanied by the smell of burning insulation which presents a very real fire hazard. Be especially careful if any fuse or other safety device is by-passed while checking.

If the blower fan operates, but the compressor clutch doesn't engage, the problem is located in the electrical circuit between the clutch coil and the main circuit. Control switches are installed in the clutch circuit to disengage the clutch to control temperature, to prevent damage and promote safety.

It is often possible to check for the flow of electricity to each of the controls and to locate at which component flow stops. Some units, such as thermal melt fuses must be renewed, but often the overriding problem such as low refrigerant charge must be correct or the new unit will quickly fail.

39. **NOISE.** Unusual noises are often an early warning of impending failure. Locating the problem and correcting the cause quickly may prevent extensive damage. Noise such as belt too loose, compressor mounting loose, bad bearing, etc., can lead to additional damage if not quickly corrected.

Some noises can be easily identified and quickly located, while others, especially intermittent noises, may be very difficult to isolate.

High refrigerant temperature or too much refrigerant will cause high pressure which sometimes expresses itself as a squealing (slipping) belt or as a

Paragraph 40   AIR CONDITIONING

Fig. 5-4 — Approved method of cleaning filters will depend upon type of construction, but all should be kept clean.

Fig. 5-5 — Electrical components are protected by fuses or circuit breakers.

loud knocking noise from the compressor.

40. NOT COOLING. Seemingly normal operation except for lack of cooling could be caused by low refrigerant charge, but do not fail to check other less obvious causes such as the heater ON and warming the air before the air conditioner evaporator cools it.

Feel the temperature of the system components. All of the lines and components on the high pressure side should feel warm, and all components on the low pressure side should feel cool. An abrupt temperature change indicates a restriction or obstruction. The inlet and outlet of the receiver-drier should be the same temperature. If the inlet is warmer than the outlet, the receiver-drier is partially plugged.

On models so equipped, observe the sight glass located in the liquid line between the condenser and the expansion orifice. If bubbles or foam can be seen, system has insufficient refrigerant. Over an extended period of time, slight loss of refrigerant is normal. Extended periods without use will permit lubricant to drain away from seals and refrigerant leakage is inevitable.

Moisture is the number 1 enemy of air conditioning. Any moisture, even very small amounts, combined with the refrigerant in the system will permit a chemical action called hydrolyzing. Hydrolysis will result in corrosion which may affect all metal parts of the system. Small particles of metal and corrosion can interfere with operation by blocking or restricting flow of refrigerant in the system. Screens and filters are installed at many locations within the system to catch the foreign particles without causing serious damage, but remember that the trouble was initially caused by permitting dirt or probably moisture to be in the system. A desiccant is included in the receiver-drier or accumulator to absorb small amounts of moisture, but the system should be evacuated using a vacuum pump to remove all of the moisture that is possible. A receiver-drier, that has absorbed all of the moisture that it can, will permit remaining moisture to circulate freely within the system. If the system operates satisfactorily during cool times of the day, but becomes intermittent when warm,

Fig. 5-6 — The sight glass can be a valuable aid in checking system condition.

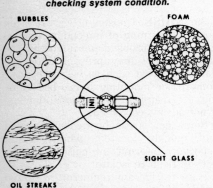

Fig. 5-7 — Only clear refrigerant should be visible through sight glass when operating properly.

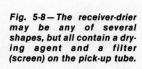

Fig. 5-8 — The receiver-drier may be any of several shapes, but all contain a drying agent and a filter (screen) on the pick-up tube.

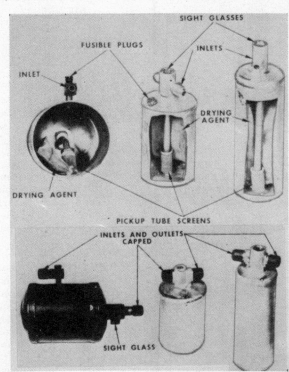

SERVICE

the desiccant may be saturated and releasing moisture during warmer temperatures. Moisture in the system can freeze in the cold sections of the system such as the expansion valve and evaporator.

An accurate temperature check of the air that has passed through the evaporator and is discharged into the cab should be made by locating a thermometer near the evaporator outlet. When the ambient temperature is above 70 degrees F. the temperature of air discharged near the evaporator should be about 45-50 degrees F. The temperature control should de-energize the compressor clutch when discharge air temperature is lower than 35-40 degrees F. and should energize clutch when temperature raises 5-12 degrees F. above the cut-out temperature. Ice will build up on the fins of evaporator if the temperature control does not de-engerize clutch when temperature falls below 35 degrees F. The compressor will cycle off then on (short-cycle) if the cut-in to cut-out temperature differential is less than 5 degrees F. Check for faulty temperature control switch or high pressure safety switch if the compressor cuts out above 40 degrees F. A jumper wire can be used to by-pass each switch to determine which is at fault, after checking to be sure that pressure is correct.

REFRIGERANT LEAK TEST

41. There are several methods for finding leaks and each should be considered in terms of safety and ability as well as initial cost. All require a refrigerant charge or partial charge (at least 50 psi with system not operating) before checking and care must be exercised to prevent injury from the escaping refrigerant. The ambient air temperature should be at least 55 degrees F. when checking.

42. **SOAP SOLUTION.** A thick solution of soap and water can be applied to a suspected area with a brush. A leak will be indicated by escaping gas creating bubbles. Some prepared solutions are also available that can be applied a variety of ways including squirting the solution directly from the container.

43. **DETECTOR TORCH.** Leak detecting torches are available in several sizes and styles. The detectors consist of a valve, a burner, propane or butane fuel cylinder and a search hose. Air is normally drawn through the search hose into the flame, but color of the flame changes when a non-combustible gas such as R-12 is drawn through the search hose into the flame. Toxic phosgene gas is produced by these leak detectors when refrigerant is passed through the flame; so be careful when operating in a closed area. The danger of an open, portable flame around any flammable material and fuel cannot be stressed enough; but the search hose can draw fuel vapor, dust and other combustible material which could result in explosion or fire. In addition to the danger involved in the operation, sensitivity is limited, making small leaks difficult to locate and operation outside in windy conditions nearly impossible.

44. **ELECTRONIC DETECTOR.** Electronic leak detectors are probably the most popular in spite of the higher initial cost. In addition to being very safe, many of these units can detect leaks as small as ½ ounce of refrigerant per year (some as small as 0.1 ounce per year) and operation is possible outside with usual amounts of wind. Most electronic leak detectors use an audible signal to announce the presence of a leak. The audible signal permits easy visual reference to the location of the probe, making operation around operating equipment safer than while watching a flame. The ability to sense a leak, even a small leak, faster, can often offset the initial cost by the time saved over less sensitive detectors.

Fig. 5-9—Drawing of typical leak detecting torch.

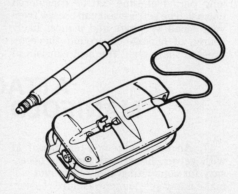

Fig. 5-11—Drawing of one type of electronic leak detector. Many different very good electronic detectors are available.

Fig. 5-10—Refrigerant fumes drawn into flame of leak detecting torch will change color of flame.

Fig. 5-12—Most electronic leak detectors can be used to sense even small leaks under windy conditions.

AIR CONDITIONING

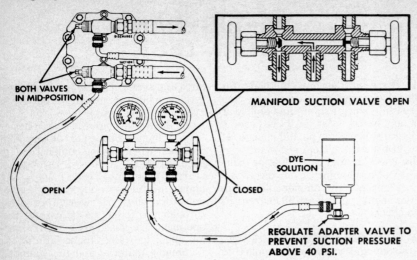

Fig. 5-13 — A dye solution with refrigerant can be used to pressurize the system and the dye will escape from leak.

45. DYE SOLUTION. Dye solution is available which can be used to charge the system when checking. The procedure, described on container, for charging the system with dye should be closely followed. The dye is contained in the liquid refrigerant and container should be inverted when filling to be sure that dye enters system. It is important to make sure that pressure does not rise above 40 psi on suction side of compressor to prevent liquid locking compressor. The dye can be seen seeping through any cracks or loose connections.

46. EVALUATION. Some leakage is normal. Small losss of refrigerant over a long period of time can be considered normal. The refrigerant can escape from very small openings and a most likely source of leakage is around the compressor shaft seal. The seal consists of two highly polished surfaces, but sealing depends upon a thin oil film filling the imperfections between the two sealing surfaces. During normal operation, oil is circulated with the refrigerant and coats the seal. Long periods without operating the air conditioner, such as during the winter, will allow oil to drain from the sealing surfaces and refrigerant will leak. Operating the air conditioner to circulate the oil and filling the system with refrigerant is often all that is necessary.

A leak in the low pressure section (evaporator, equalizer line for expansion valve, compressor crankcase and line between evaporator and compressor inlet) will leak more slowly than a similar leak in the high pressure section. Leaks in the high pressure section are usually large, sudden, result in loss of all refrigerant and can be seen without needing a leak detector. The most likely sources of all leaks are connections, gaskets, seals, unsupported ridged tubing, nicks in evaporator and corroded condenser coils. Remember that R-12 is heavier than air and the gas can be detected well below the leak source. Also be sure that the detector is not detecting leakage from a refrigerant service container or other equipment in the area. The flame will also change color on leak detecting torches if dust, dirt, gasoline, etc., is drawn into the flame.

ATTACHING MANIFOLD AND LINES

47. An air conditioner test manifold with valves, hoses and gages is necessary for conducting many tests and will make many servicing procedures easier. The blue hose should be attached to the low side of manifold, red hose to high pressure side and the yellow hose should be in center service connection of manifold. The end of hoses not connected to manifold usually have a projection which

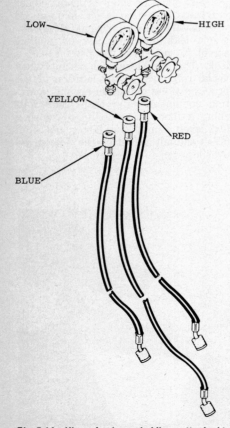

Fig. 5-14 — View of color coded lines attached to manifold and gage assembly.

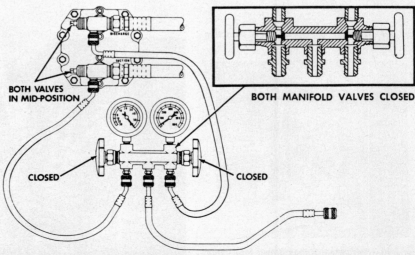

Fig. 5-15 — The red hose should be attached to high pressure service port and blue hose should be attached to low pressure service port.

SERVICE

Paragraph 47 Cont.

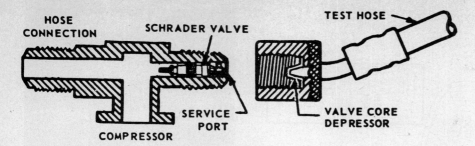

Fig. 5-16—Some service ports are equipped with automatic Schrader type valves.

automatically depresses the Schrader valve in service port. Many manifold assemblies have storage attachment points for free ends of the hoses so that dirt, moisture, etc., will not contaminate hoses.

To attach the manifold pressure gage set to the system for service, make sure that valves on manifold are closed and put on safety glasses or goggles. Identify the suction port of compressor, remove protective cover, then attach the blue low pressure hose to the service port. Usually the suction side of compressor is marked, but the line from evaporator is attached at this point. Be sure that the connector for the low pressure hose is tightened firmly by hand, but do not damage parts by overtightening. Attach the red high pressure hose, similarly, to other service port on compressor. The high pressure compressor outlet may be marked "DISCHARGE" and the line connected to this port runs directly to the condenser. Be careful when attaching hoses to service ports because some refrigerant may escape around connector.

NOTE: Service ports may sometimes be located on lines or components instead of at compressor. Be sure that low and high pressure gage lines are correctly attached. On some models the high pressure connector is smaller to ensure correct attachment of test and fill hoses.

Two different types of valves are used to keep the service port closed during normal operation when the test manifold and hoses are not installed. A Schrader valve is installed in service port of many systems and a projection in hose connection depresses valve automatically opening it when the hose is connected. A second type of valve must be manually opened by turning the valve spindles. The manual valves have three positions as shown in Fig. 5-17. During normal operation, the valve is back-seated as shown in view "A," closing the service port and opening the passage between compressor and hose port. The mid-position as shown at "B" is used to open the service port to the operating system. Front seating valve as shown at "C" isolates the compressor from the hoses and remainder of system.

Open the service port on manually opening valves by turning valve to the mid-position.

When a hose or any other part is attached to the system, the new part should have the air removed by filling the part with refrigerant. This is called **"PURGING"**. When the manifold assembly is attached, purge these parts by opening low pressure valve of manifold slightly, allowing some refrigerant to escape from yellow hose (attached to center of manifold), close valve, then purge high pressure side similarly.

CAUTION: Use care to release refrigerant into safe area such as shop exhaust collector. Do not release more refrigerant than necessary and release refrigerant slowly.

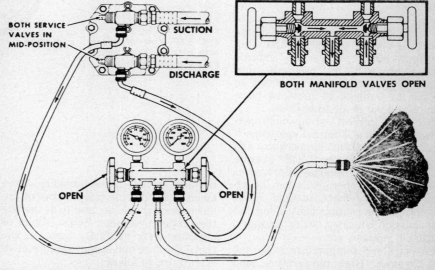

Fig. 5-18—Purge air from manifold and lines after attaching to prevent air from being introduced into system.

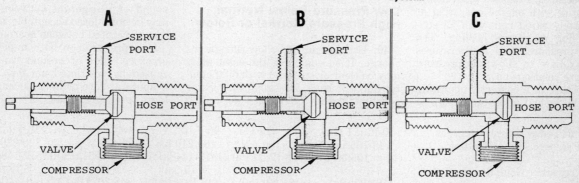

Fig. 5-17—Manual service valves are installed on compressor. View "A" shows back-seated, with service port closed; compressor port is open to the hose port. View "B" shows valve in mid-position which provides open passage and connection of all three ports. View "C" shows valve in front seat position which closes hose port and can be used to isolate compressor.

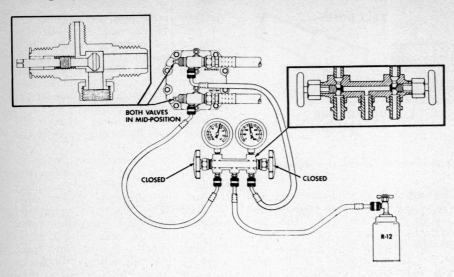
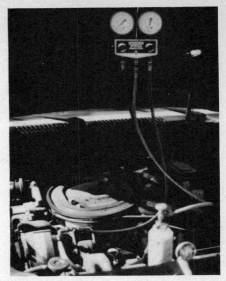

Fig. 5-19 — View showing typical connections for checking system pressure.

SYSTEM PRESSURE TEST

48. The pressures developed in the high pressure and low pressure sides of system indicate whether or not the system is operating properly and often the incorrect pressures will indicate what is wrong. To test pressures, first connect a manifold and gage set to system as outlined in previous paragraphs and purge air from lines.

The exact "correct" or "normal" operating pressures will depend upon capacity of system, ambient temperature and humidity. The system will operate satisfactorily over a range of pressures and ambient temperatures can be affected by very hot localized areas resulting from routing lines along hot transmission housings, water passages, warm air streams, etc.

Pressure and temperature raise and fall together. High temperature (and high humidity) will result in higher system pressures. Pressure tests should be conducted with engine operating at approximately rated rpm with all controls set for maximum cooling. The following pressures and temperatures are approximate to show the pressure temperature relationship.

The pressure in a correctly operating system will depend upon the design of the unit as well as the humidity of the ambient air; but the temperature of the ambient air should be used to determine the proper pressures. Some results of pressure tests on faulty systems, the probable cause and suggested corrective actions are as follows:

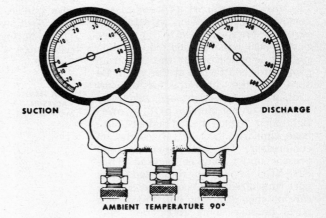

Fig. 5-20 — View of gages with both pressures below normal.

Low Pressure Below Normal, High Pressure Normal or Below

49. Usually caused by low refrigerant charge. If so equipped, the sight glass may be clear or streaked with oil if empty; but will have occasional bubbles or constant stream of bubbles depending upon degree of discharge. The low pressure gage will indicate a vacuum if empty, but will show varying degrees of pressure depending upon amount of refrigerant remaining in the system. Extremely low pressures usually indicate a leak in the system that must be located and repaired before system will hold charge; however, loss of ½ to one pound of refrigerant between seasons may be considered acceptable, especially if not operated between seasons.

Low pressure on both high and low pressure sides of system can also be caused by restricted air flow through evaporator core or by a blockage in the high pressure side of refrigerant

Ambient Temperature – Degrees	70 F. (21 C.)	80 F. (27 C.)	90 F. (32 C.)	100 F. (38 C.)	110 F. (43 C.)
High Pressure – Gage Reading	145-155 psi (1000-1069 kPa)	170-180 psi (1172-1241 kPa)	200-210 psi (1379-1448 kPa)	215-225 psi (1482-1551 kPa)	220-230 psi (1517-1586 kPa)
Low Pressure – Gage Reading	7-15 psi (48-103 kPa)	7-15 psi (48-103 kPa)	7-15 psi (48-103 kPa)	7-30 psi (48-207 kPa)	7-35 psi (48-241 kPa)
Approximate Evaporator Discharge Air Temperature – Degrees	46-49 F. (7.8-9.5 C.)	47-50 F. (8.5-10.0 C.)	49-52 F. (9.5-11.1 C.)	50-53 F. (10.0-11.7 C.)	51-54 F. (10.6-12.2 C.)

SERVICE

Paragraphs 50-52

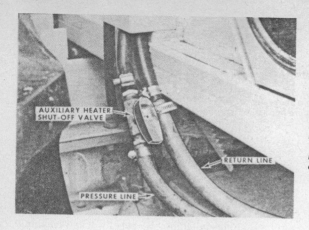

Fig. 5-21—If the heater and air conditioner are both operating, the circulating air will probably not be cool, because the heater is much more effective. The heater usually has an auxiliary shut-off to prevent operation during summer.

passage. Some possible causes of refrigerant blockage are: Screens at expansion valve or receiver-drier plugged, moisture in system freezing and stopping refrigerant flow or expansion valve damaged and not opening.

If low pressure gage reading is actually a vacuum, a complete blockage is indicated. Shut the system off and allow it to warm up to room temperature. Check the lines to be sure that all self sealing connections are attached and tight. Start engine and set system for maximum cooling. If the evaporator will now become cool, the expansion valve was frozen because of moisture in the system. Discharge refrigerant from system, renew the receiver-drier assembly, check for leaks, then evacuate and charge the system.

A new receiver-drier should be installed as a precaution against internal icing of expansion valve, if the system has been opened three times, has remained open for an extended period of time or if the condition of the drier is in any way questionable.

Check the system between the receiver-drier and the low pressure service valve for restrictions by feeling all of the connections and components. Any portion that is cold to the touch or that frosts up, indicates a restriction in the refrigerant flow.

Low Pressure Normal High Pressure Normal

50. Normal system pressure without cooling can be a result of several problems. Be sure that the heater core is not hot. Most systems have shut-off valves on or near the engine at the heater hoses to stop hot water from circulating into heater core. Air in the refrigerant lines will not provide proper cooling, but can maintain near normal pressures, especially if balanced with too much refrigerant. Gage needles will probably shake and bubbles will occasionally be present in sight glass.

If the system cools normally for a while (usually in the morning), then stops cooling, the desiccant in receiver-drier has probably reached its saturation point. Moisture will be released from desiccant later during operation at high ambient temperature and can freeze in expansion valve orifice stopping refrigerant. Corrective action is to drain refrigerant, install new (dry) receiver-drier, evacuate system, then recharge system.

Low Pressure Normal, High Pressure Above Normal

51. Complaints about compressor noise are very often caused by too much refrigerant. Cooling will usually be normal, the low pressure may be normal (may also be too high), no bubbles in sight glass, but the high pressure will be too high. The compressor may lock up (especially with loose-drive belts) and hoses may burst or blow away from connections.

Low Pressure Above Normal, High Pressure Below Normal

52. If the low and high pressures are the same, the compressor is not working. Varying degrees of compressor wear or failure will be indicated by the two gages indicating more nearly the same pressure. Isolate the compressor and check reed valves, gaskets, piston rings, etc., for wear or damage.

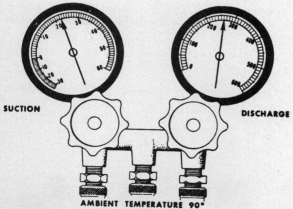

Fig. 5-22—View of gages with normal low pressure, but high pressure above normal.

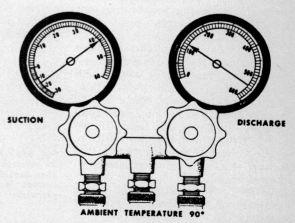

Fig. 5-23—View of gages with low pressure above normal; high pressure below normal.

Paragraphs 53-55

AIR CONDITIONING

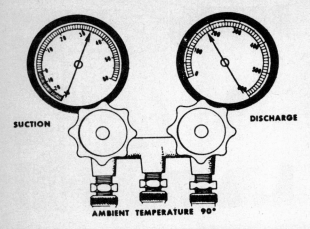

Fig. 5-24—View of gages with low pressure above normal, high pressure normal.

Fig. 5-25—Fins on condenser must be straight to radiate properly.

Low Pressure Above Normal, High Pressure Normal or Above

53. Too much refrigerant, expansion valve not opening or condenser malfunction can cause pressures to be too high.

Too much refrigerant can be installed in the system only by forcing. If both low and high pressures are above normal, the compressor is probably very noisy and hoses are approaching burst pressure. Increases in ambient temperatures will rapidly increase operating pressures above tolerable levels. Lower pressures by allowing some refrigerant to escape from service hose, while checking pressures.

Improper operation of the expansion valve can cause the valve to remain open flooding the evaporator. Liquid or high pressure refrigerant will enter the compressor probably causing compressor to be very noisy. The suction side of compressor, crankcase and head will be cooler than normal and may frost up. Check to be sure that the temperature sensing element is securely clamped to the evaporator outlet pipe and is properly covered with insulating tape. Check operation of the expansion valve by detaching the sensing element from the outlet pipe, then heating or cooling the element while checking the operating pressures. The element can be warmed by hand and should cause the low pressure to be above normal. The element can be cooled by application of ice or by carefully releasing liquid refrigerant from refilling container onto the sensing element. When the sensor is cold, the expansion valve should close and the result should be lower than normal pressure on the low pressure gage. Install new expansion valve if changing the temperature of sensor element does not change operating pressure.

Condenser malfunction could be restriction of refrigerant flow or heat build up. Restriction of flow can be caused by dented core, by an accumulation of oil in condenser core or by a plugged receiver-drier. Restriction can be located by noticing temperature drop at point of restriction, but be careful because many sections will be very hot.

The pressure and temperature raise and lower together and the purpose of the condenser is to change the refrigerant from vapor to liquid by removing some heat from the refrigerant. If heat is not removed, the refrigerant will not condense properly and the pressure will be too high. Check to be sure that condenser fans are operating correctly and that air flow is not blocked. Straighten fins on condenser if necessary using a special comb. Be careful when cleaning not to force dirt between engine radiator and condenser on models with condenser in front of cooling system radiator.

COMPONENT TESTS

54. Many components can be isolated and checked to be sure of their operation. Refer to the appropriate following paragraphs for some suggested tests of individual components.

Inadequate lubrication, plugged system or liquid entering the compressor are the most common causes of compressor damage. Often several problems will combine to ruin a compressor quickly. Lubrication is important to the compressor and some systems contain only a small amount which is circulated with the refrigerant. Anything that stops circulation of refrigerant, such as a clogged orifice, will also stop the flow of lubricant. The combination of excessive load caused by the plugged system and inadequate lubrication also caused by the plugged system will quickly ruin the compressor.

Thermostatic Control Switch

55. The thermostatic control switch should be checked to be sure that switch

Fig. 5-26—View of typical thermostatic control switches. Sensing tubes must be correct length to enter evaporator at correct location.

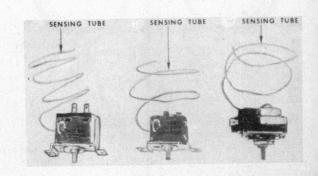

SERVICE

Paragraphs 56-57

will open or close within a temperature range of 8-12 degrees F. (4.4-6.7 degrees C.). Do not attempt to check opening or closing at specific accurate temperatures. Insert a dial type thermometer in the evaporator coil fins near the sensing tube (one fin right or left from the tube). Start engine, turn fan switch on and operate engine at 1200 rpm for 10 minutes. Turn thermostatic control switch to hottest temperature position and check clutch for engaging or disengaging with a temperature change. Turn thermostatic switch to coldest temperature setting and note clutch operation. The clutch should disengage about 32 degrees F. (0 degrees C.). If the thermostatic switch does not operate properly, install a new switch assembly.

Receiver-Drier

56. Operate the air conditioner for about five minutes, then slowly feel the lines and the receiver-drier to detect changes in temperature. Be careful to prevent injury. There should be no noticeable difference in temperature. If cold spots are felt, it indicates that the unit contains moisture and/or flow is restricted. In either case, the receiver-drier must be renewed. Check the new receiver-drier to be sure that all protective caps are tightly in place. New receiver-driers can absorb their capacity of moisture before installation in system, if not kept tightly closed while in storage.

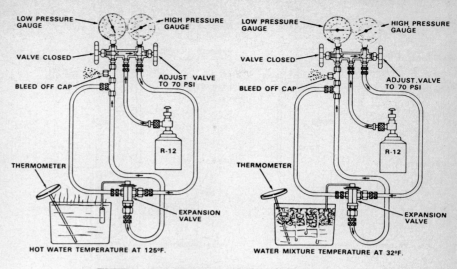

EXTERNALLY EQUALIZED EXPANSION VALVE

Fig. 5-28 — Refer to text for checking externally equalized expansion valves.

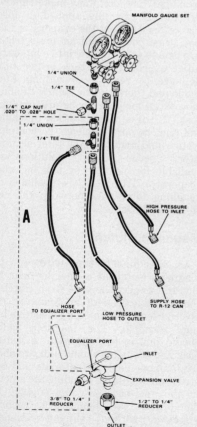

Fig. 5-27 — Test connections for checking expansion valve. Parts shown inside "A" are necessary only for externally equalized units.

Expansion Valve

57. Before removing valve from system, attach the manifold gage set and run engine at about 1000 rpm with air conditioner controls set for maximum cooling. Stop the engine, if low pressure does not stabilize (varies more than 10 psi, 69 kPa) or if pressure remains above 50 psi (345 kPa) and compressor knocks. Detach the expansion valve thermal bulb from the evaporator outlet pipe and clean the bulb, clamp and outlet pipe thoroughly. Reinstall bulb, clamping it securely to the evaporator outlet tube, then insulate the tube and bulb by wrapping with insulating tape.

Recheck operation, but stop engine if previous condition exists or if low pressure drops below zero. Exhaust the refrigerant charge slowly through the low pressure valve of manifold and observe the high pressure gage. If high pressure does not drop below 70 psi (483 kPa), the expansion valve is stuck shut and must be renewed (or disassembled and repaired). If the high pressure drops, exhaust the entire charge, evacuate the system and refill with the correct amount of refrigerant. If the system now functions normally, the problem was caused by a low charge or moisture. Perform leak test and repair if necessary.

Most expansion valves can be disassembled, cleaned, reassembled, tested and adjusted. Usually installation of a new valve provides a much more satisfactory repair for expansion valve problems, but testing the valve while removed may be desirable. Visually check and discard the expansion valve if the tube from the temperature sensing bulb, equalizer tube (if so equipped) or valve body is damaged enough to leak or impair operation. Equalizer tube and inlet screen filter should be open and clean.

Two tests can be made; one for maximum flow and one for minimum flow. The test results may vary from those listed because of the difficulty in maintaining accurate temperatures and pressures, but the data should be sufficient to determine if valve is operational. Use the necessary "T" fittings, unions and reducers (Fig. 5-27) to provide the test connections as shown in Fig. 5-28 or Fig. 5-29. The bleed off orifice is a 0.020-0.028 inch (0.51-0.71 mm) hole drilled in the center of a cap nut.

For maximum flow test, insert the sensing element (bulb) in water that is 125 degrees F. (52 degrees C.). Open the high pressure valve enough to cause the high pressure gage to read 70 psi (483 kPa). The low pressure gage should read 43-55 psi (296-379 kPa). Pressure over 55 psi (379 kPa) indicates flooding and pressure under 43 psi (296 kPa) indicates valve is not opening.

Prepare a container of water with as many ice cubes as possible for checking minimum flow. Insert the expansion valve sensing element into the ice water and allow time for the valve to react to

AIR CONDITIONING

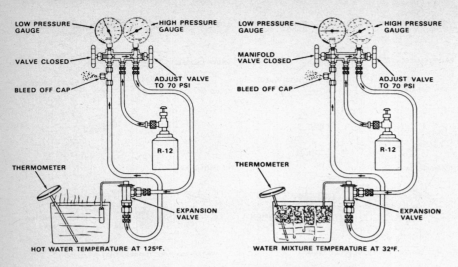

INTERNALLY EQUALIZED EXPANSION VALVE

Fig. 5-29 — Refer to text for checking internally equalized expansion valves.

the near 32 degrees F. (0 degree C.) temperatures of the ice water. Open the high pressure valve enough to cause the high pressure gage to read 70 psi (483 kPa). The low pressure gage should read about 19½-22½ psi (134-155 kPa); but exact pressure will depend upon which superheat spring is used in the expansion valve. Refer to the specifications that follow:

Superheat Setting Fahrenheit (Celsius)	Low Gage psi	Pressure (kPa)
5 (-15)	23-26	(159-179)
6 (-14.4)	22¼-25¼	(153-174)
7 (-14)	21½-24½	(148-169)
8 (-13.3)	21-24	(145-165)
9 (-12.8)	20¼-23¼	(140-160)
10 (-12)	19½-22½	(134-155)
11 (-11.7)	19-22	(131-152)
12 (-11)	18-21	(124-145)
13 (-10.6)	17½-20½	(121-141)
14 (-10)	17-20	(117-138)
15 (-9.4)	15½-18½	(107-128)

Cleaning may be attempted, but be sure to locate original setting of the superheat adjusting screw before disassembling valve. Refer to Fig. 5-30 for typical cross-section and exploded views. Clean parts in mineral spirits, drain, then blow dry with refrigerant-12. Lubricate parts with special refrigerant oil when assembling, then recheck maximum and minimum flow. Turning the adjusting screw one complete turn will change superheat setting approximately 3-5 degrees F. (2-3 degrees C.). Discard expansion valve if valve does not respond correctly.

Orifice Tube

57A. The orifice tube is located in the inlet to the evaporator and provides a hole about 1.78 mm (0.070 inch) diameter to limit the flow of liquid refrigerant into the evaporator. Cycling Clutch Orifice Tube (CCOT) systems do not usually incorporate a filter or drier between the compressor and the evaporator, so problems caused by moisture are more prevalent with this type. Moisture or resultant corrosion and sludge will partially or completely block the orifice. Some air conditioning service shops believe the orifice tube should be removed and visually inspected and cleaned each time system is opened for other service.

To check for blockage, attach the manifold and lines to the system as described in paragraph 47. Check temperature of the evaporator outlet tube, accumulator and evaporator inlet tube. If suction (low side) pressure is too low and the temperature of the accumulator and evaporator outlet is warmer than the evaporator inlet, attempt to add refrigerant as described in paragraph 63. If the accumulator and evaporator outlet cools to the same temperature as the evaporator inlet, refrigerant charge was just low. If, however, the accumulator remains warm, pressure remains low and a ring or band of frost is around the evaporator inlet pipe, the orifice tube is clogged and should be serviced.

The orifice tube is inserted into the evaporator inlet pipe and located by two small dimples in the pipe. Special removal tools are available for withdrawing the orifice tube; however, built up sludge, dirt, etc., may make removal difficult. If carefully applied, a torch may be used to heat the evaporator inlet pipe, then compressed air (or refrigerant) can be used to blow into evaporator outlet expelling the stuck orifice tube. Be sure to remove everything flammable and thermostatic switch capillary tube from evaporator before applying heat. If orifice tube is stuck by sludge and dirt, be sure to clean the complete system and install a new filter-accumulator. Special flushing solutions are available for cleaning system including evaporator. Also, kits are available for installing filter-drier between the condenser and evaporator.

Magnetic Clutch

58. Disconnect the wire to the magnetic clutch at connector near the clutch. Connect the wire from clutch to the negative lead of an ammeter and positive lead of ammeter to positive battery terminal. The magnetic clutch should pull in (engage) with a distinct click and ammeter should indicate about 3-3.5 amps. Various materials and construction details may change the amount of current necessary to engage the clutch, but all should lock securely engaged, without slippage. Usually clutch disengaged clearance is adjustable and, if correctly set, the clutch will disengage as soon as electrical connection is broken.

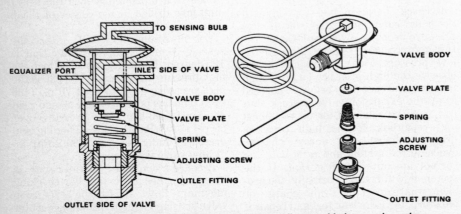

Fig. 5-30 — Drawing showing typical cross-section and disassembled expansion valve.

MAINTENANCE

The air conditioner will give better service if everything is kept clean, belts are tight, correct amount of refrigerant is maintained in system, any problems relating to the system are quickly repaired and the system is operated often enough to keep lubricant circulated.

59. ROUTINE SERVICE. The importance of cleanliness cannot be overstressed. Nearly all systems used on farm equipment have filters to clean the air entering the cab, but operation will be affected if these filters are permitted to stay dirty. Air conditioning operates by moving heat from inside the cab to outside, but this can't be done efficiently with layers of dirt on the system components. Dirt will block the flow of air and will insulate components. Dirt on the evaporator will keep the refrigerant from absorbing heat from inside the cab. Dirt on the condenser will prevent radiation of heat to the outside air which will result in both high temperature and high operating pressure. Do not use steam to clean parts of the air conditioning system which are charged with R-12. The high temperature could cause enough pressure to burst part of the system.

The compressor drive belts should be kept tight and aligned. Drumming noise, vibration, etc., are often caused by the condition or tightness of compressor drive belts. Check for refrigerant overcharge and misalignment of pulleys if drive belt problems are encountered.

If so equipped, check the sight gage occasionally with system operating to be sure the system remains fully charged. Bubbles flowing past the sight glass indicates low refrigerant level. The system should be checked for leaks and filled with proper amount of refrigerant. Slight leakage over a long period of time is normal and system can be filled with refrigerant.

Do not let the air conditioner remain broken. Keep the system in good working order and operate the system occasionally even during cool weather and when the equipment would normally sit idle. Operating the system will circulate the lubricant and may keep the seals from drying out and leaking refrigerant. The extent of damage that will be caused by allowing the system to remain broken will depend upon the original failure. A simple leaking hose or connection will permit the refrigerant to escape and air (with moisture) will enter the system if allowed to set. The moisture can corrode internal metal parts and contaminate the receiver-drier.

During maintenance or repair, use only refrigerant oil on the threaded connections. Do not use any other materials as thread compounds. When adding refrigerant to the system, use only Refrigerant-12. Before the system can be opened for renewal of lines or components, the system must be completely discharged. Whenever the system has been opened, it must be swept with a partial charge and the entire system tested for leaks. When the system is opened, the compressor oil level should be checked and oil added if necessary. A new receiver-drier should be installed if the condition of old unit is in anyway questioned. Most manufacturers suggest that receiver-drier be renewed after opening system three times for any reason, but it may be difficult to account for previous repair.

Fig. 6-2 — Manifold (2) and valves (3) connected to service ports (1) can be used to safely discharge system into shop exhaust (4).

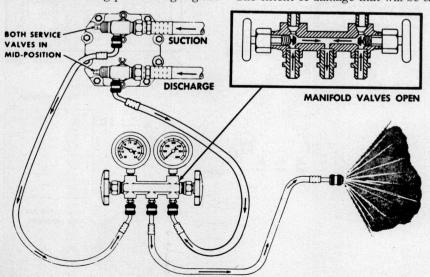

Fig. 6-1 — Be sure to discharge system safely into well ventilated area.

DISCHARGING SYSTEM

60. Attach the manifold gage set and attach a long hose to the center connection of the manifold. Place other end of

AIR CONDITIONING

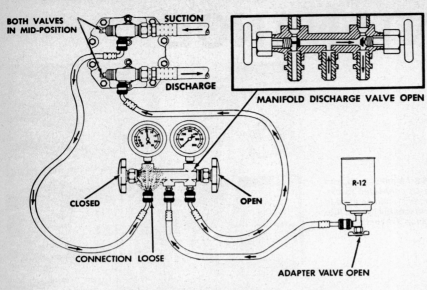

Fig. 6-3 — The system except compressor can be swept with refrigerant as shown to remove most air from the system.

CAUTION: Be extremely careful to never start compressor (or even engine) with high pressure valve open. Turn container upside down to fill with liquid refrigerant only through high pressure service port and only with compressor stopped and engine not running.

A leak test should be accomplished each time sweep-test charge is installed in system. The sweep-test charge is necessary for leak checking system which has accidently lost its charge and a complete leak test should be performed after repair has been completed to be sure that fittings are tight, that repair is successful and that no other leaks are present in system.

If any parts of the refrigerant system were disconnected, discharge the sweep-test charge after checking for leaks, evacuate the system, then fill system as outlined in the appropriate following paragraphs 62 and 63.

this hose into an exhaust ventilation system outlet or to the outside of the building. Be sure that both manifold valves are closed. Open windows, operate engine at high idle, turn temperature control to coldest position and fan switch to high, then allow system to operate at full capacity for 15 minutes. **This will cause most of the compressor oil in the system to return to the compressor crankcase.** Stop engine and open high pressure manifold valve only a small amount to permit the refrigerant to discharge slowly. Allow system to discharge until the high pressure gage registers zero, then open the low pressure manifold valve to release any vapor trapped in the suction side of system.

SWEEP-TEST CHARGE

61. The purpose of the sweep-test charge is to pressurize the system so that a leak test can be made. The sweep-test charge also serves the purpose of drying the system or sweeping out trapped moisture such as when new component is installed.

CAUTION: The compressor will be extensively damaged if operated without proper lubrication. The only oil lubricating some (including Delco Air R-4, GM lightweight DA-6 and the GM Harrison 5 cylinder models) compressors is the oil circulated with the refrigerant in the system. The refrigerant used to sweep charge is dry (contains no oil), so before operating the compressor, be sure that adequate lubricant is present within the complete system and lines. An oil reservoir is part of the accumulator of systems with this type of compressor to control the flow of oil.

To install sweep-test charge, first attach manifold and lines to the system as described in previous paragraph 47. Connect dispensing valve and container of refrigerant to the center (yellow) hose of manifold. Open the dispensing valve, turn can of refrigerant upside down (for liquid) and open the high pressure valve of manifold. Loosen the blue low pressure hose connection at manifold to release air until refrigerant is escaping, then tighten hose connection and close high pressure valve.

EVACUATING SYSTEM

62. The purpose of evacuating the system is to remove moisture. **Lowering the pressure to a vacuum lowers the boiling point of water.** Many different types of vacuum pumps are available but a good pump should be able to create a vacuum sufficient to cause water to boil (vaporize) at about 65 degrees F. (18.3 degrees C.). Evacuation should only be accomplished when ambient temperatures are warm. Effectiveness of evacuation is dependent upon the amount of vacuum (to lower

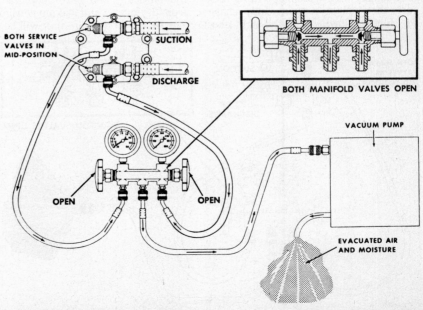

Fig. 6-4 — To use a vacuum pump will help remove moisture from the system by lowering the boiling point.

SERVICE

Paragraph 63

boiling point), the ambient temperature (to provide heat for boiling) and the length of time (to permit enough time for moisture to boil out).

To evacuate the system, attach manifold and gages to the system and discharge all refrigerant from system. Attach center (yellow) service hose to the vacuum pump, start pump and open low side manifold valve. The low pressure manifold gage should immediately indicate a lowering of the pressure and should indicate a vacuum of about 28-29 inches of Mercury (71-74 cm of Hg) after five minutes. If low pressure gage does not steadily lower, close the manifold valve and observe gage. If low pressure rises, check for leak in system. Check to be sure that the high pressure gage indicates less than zero pressure. If high pressure gage needle does not drop to zero or below, discontinue evacuation and check for system restriction. If high pressure gage lowers to zero or below, open the high pressure valve for faster more complete evacuation. Usually the system should be evacuated for at least 30 minutes at the lowest attainable pressure; however, especially wet and/or contaminated systems require additional treatment.

If system is open for a long period of time, opened in such a way to allow entry of water or otherwise contaminated, observe the following procedures. Discharge all gas from the system, drain all of the oil that is possible, then purge system with pressurized refrigerant. Do not use compressed air. Connect the vacuum pump to the system for about 30 minutes after attaining lowest possible reading while heating metal components of the system to about 150 degrees F. (65.6 degrees C.) to drive moisture out of system. Close valves, disconnect vacuum pump, then add a sweep-test charge as outlined in a preceding paragraph. After about five minutes, release the refrigerant quickly to help draw more contaminates from the system. Refer to LUBRICATION paragraphs and add the correct amount of oil to the system. Evacuate system again leaving vacuum pump operating for about 30 minutes after lowest attainable gage pressure has been reached. Again add sweep-charge to system and operate air conditioner at maximum cooling for about 3 minutes, then stop operation of the unit and discharge refrigerant quickly. Check oil level and fill to correct amount. Remove old receiver-drier assembly and install new unit. Purge system with refrigerant as outlined in paragraph 61, discharge the sweep-charge, then evacuate system to maximum vacuum for about 30 minutes.

After system is evacuated, it may be possible to fill with refrigerant using the existing vacuum to draw the initial charge into system without operating the compressor. Refer to the following paragraph 63 for FILLING SYSTEM.

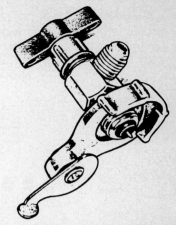

Fig. 6-6 — View of a typical dispensing valve. This type of valve attaches to top of container.

FILLING (CHARGING) SYSTEM

63. The system should be filled only after leak testing to determine cause of low refrigerant charge. Slow leakage may be considered normal, but check to be sure that fittings, front seal of compressor or some other area is not leaking. Small leaks often grow into larger ones if permitted to go unchecked. Check the complete system after any repair to be sure that there are no other small leaks.

The refrigerant can be filled with or without the manifold, gages and valves. The preferred (and easiest) method uses the manifold and gages. The system should be swept and evacuated if completely discharged to assure that all air and moisture are removed before filling.

Proceed as follows when using manifold and gages: Refer to preceding paragraph 47 for details of attaching manifold and lines to system. Purge manifold and lines slowly and safely. Attach a dispensing valve to container of refrigerant and connect valve to the center (yellow) hose of manifold. Open dispensing valve, then loosen center hose connection at manifold to purge air and fill the center hose with refrigerant.

CAUTION: The compressor will be extensively damaged if operated without proper lubrication. The only oil lubricating some (including Delco Air R-4, GM lightweight DA-6 and the GM Harrison 5 cylinder models) compressors is the oil circulated with the refrigerant in the system. The refrigerant used is dry (con-

Fig. 6-5 — View showing typical vacuum pump attached to manifold and gage set. Service ports are at (1), low pressure gage at (2), high pressure gage at (3), valve handles at (4), vacuum pump at (5).

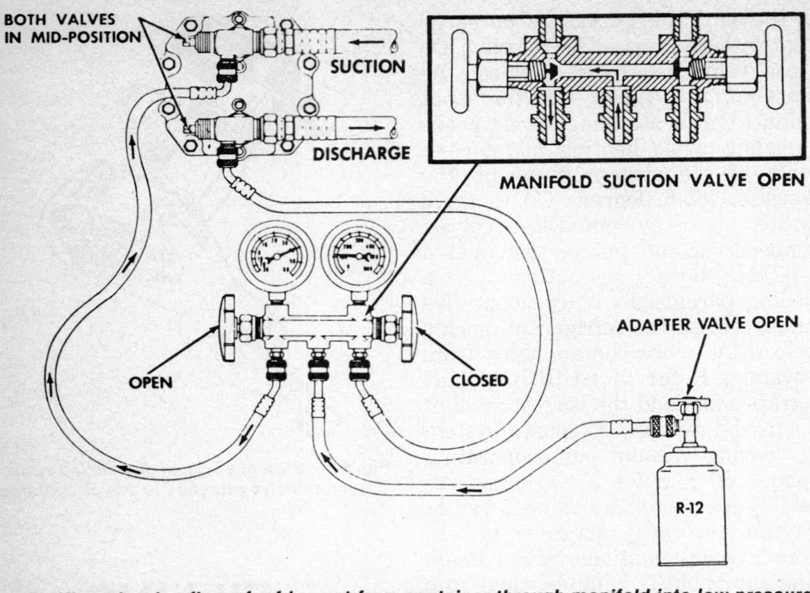

Paragraph 63 Cont.

AIR CONDITIONING

Fig. 6-7—View showing flow of refrigerant from container through manifold into low pressure (suction) service port.

tains no oil), so before operating the compressor, be sure that adequate lubricant is present within the complete system and lines. An oil reservoir is part of the accumulator of systems with this type of compressor to control the flow of oil.

If pressure is less than 40 psi (276 kPa) with **air conditioner not operating**, open the high pressure valve enough to maintain a steady pressure on gage not exceeding 40 psi (276 kPa). Refrigerant leaving container through the center service hose will cool the container and the cool temperature will reduce pressure in the can. Temperature of the refrigerant in the dispenser containers can be maintained by heating can in pan of hot water. Do not allow can to exceed 125 degrees F. (51.7 degrees C.), do not use an open flame to heat container. After filling and before starting engine, **close the high pressure valve of manifold and be sure refrigerant container is upright.**

Refrigerant capacity of most systems is about 50 to 60 fluid ounces and much of required refrigerant can be filled without starting engine or engaging compressor. It will often be possible to fill system without engaging compressor if the system has been thoroughly evacuated.

CAUTION: Be extremely careful to never start compressor (or even engine) with high pressure valve open. Container may be turned upside down to fill with liquid refrigerant only through high pressure service port and only with compressor stopped and engine not running.

To fill a partially filled system, be sure that high pressure valve on manifold is closed and that refrigerant container is upright. Start engine and run at about 1500-1700 rpm, then set air conditioner controls for maximum cooling.

NOTE: It should not usually be necessary to by-pass the low pressure safety switch to operate compressor. Initial charge with unit not operating should be sufficient to engage the low pressure safety switch.

Open the low pressure valve, make sure refrigerant container is upright and allow vapor to flow from can, through the service hose and into low pressure side of system. Watch the sight glass, the low pressure gage and the high pressure gage while filling. The bubbles will disappear from the sight glass and pressures will raise to acceptable levels when system is full and operating properly. Be careful not to overfill system.

Correct pressures for filled system will depend upon condition of compressor, capacity of system and design of the system, but the most important variable is temperature. If specific recommended pressures are not available for the system being filled, use the following standard pressures depending upon ambient air temperature.

AMBIENT AIR TEMPERATURE		HIGH (DISCHARGE) PRESSURE		LOW (SUCTION) PRESSURE	
F.	(C.)	PSI	(kPa)	PSI	(kPa)
70°	(21°)	145-155	(1000-1069)	7-15	(48-103)
80°	(27°)	170-180	(1172-1241)	7-15	(48-103)
90°	(32°)	200-210	(1379-1448)	7-15	(48-103)
100°	(38°)	215-225	(1482-1551)	7-30	(48-207)
110°	(43°)	220-230	(1517-1586)	7-35	(48-241)

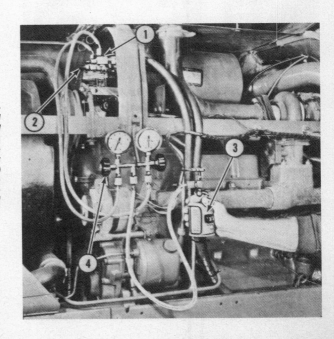

Fig. 6-8—View showing connections for filling typical system. Hoses are attached to correct service ports (1), manual valves (2) are opened, if so equipped, low pressure manifold valve (4) is opened to permit refrigerant from container (3) to enter low pressure side of system.

SERVICE

Paragraph 64

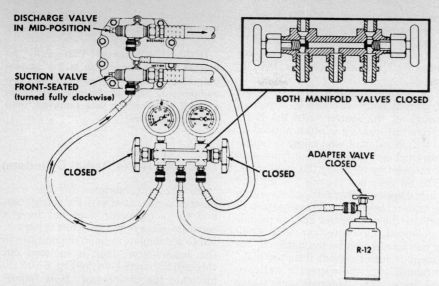

Fig. 6-9—Manual valves on compressor can be used to isolate system as described in text if so equipped.

The air conditioner can be similarly filled through the low pressure service port using only a servicing hose and a dispensing valve. Attach dispensing valve to container, hose to valve and hose to low pressure service port. Purge air from hose and valve by opening the dispensing valve slightly while attaching hose to service port. If service port has manual valve, open valve to mid-position. The service hose must have valve depressor in connection at service port end if service port has Schrader valve. Start engine and operate air conditioner at maximum cool. Open dispensing valve and observe sight glass. Close valves when bubbles disappear from sight glass. Close manual service valve if so equipped, then detach hose from service port.

CAUTION: Do not overfill system and do not turn refrigerant container upside down. Either of these actions can cause excessive pressure which can result in damage and/or injury. Be sure to purge service hose or the column of air contained in hose will be forced into system contaminating the refrigerant.

The refrigerant container can be placed in container of warm water not hotter than 125 degrees F. (51.7 degrees C.) to increase pressure in container and speed flow of refrigerant into system.

If one container is not enough, close manual service valve, if so equipped, or disconnect hose from service port if equipped with Schrader valve. Remove empty container, attach dispensing valve to full container, purge service line, then reconnect hose and continue filling procedure.

Always install covers over service ports when finished charging system.

ISOLATING AND PURGING COMPRESSOR

64. The compressor can be isolated from remainder of the system on models equipped with manual type servicing valves. Attach manifold and gage set to the high and low pressure service ports as described in paragraph 47, operate air conditioner at maximum cooling for about 10-15 minutes with engine running at 1500 rpm. Slowly close (front seat) the low pressure (suction) side service valve until the low pressure gage indicates zero, or slight vacuum. Slow engine to idle speed, stop engine, then completely close (front seat) the high pressure service valve by turning fully clockwise. Open the manifold valve nearest the high pressure gage to allow refrigerant trapped inside compressor to escape. Release the refrigerant slowly and safely in a well ventilated area. The compressor is isolated from the remainder of the system permitting service to compressor and very little of the refrigerant charge is lost. Leave hoses attached to service valves and detach service valves from compressor if compressor is to be removed. The hoses, evaporator, condenser and receiver-drier will remain charged with refrigerant.

Models **without manual service valves** must have complete system discharged as outlined in previous paragraph 60 to safely service compressor. Some models have self-closing quick disconnect fittings on the hose between compressor and condenser and on hose between evaporator and compressor. If equipped with self-closing fittings, the compressor can be isolated by separating hoses at these connections. Be sure to release refrigerant trapped in compressor and both lines before attempting any service. The manifold, gage and valve set can be used to safely release refrigerant from compressor through the service hoses.

On all models, be sure to use new seal rings or gaskets when reconnecting service valves to compressor. Tighten all screws, nut and "Rotalocks" to the specified torque when assembling. The following torques are for installing service valves to compressor.

DELCO AIR (6 Cylinder)
 Suction and discharge
   connector screw.........25 ft.-lbs.
                           33.9 N·m
TECUMSEH (2 Cylinder)
 "Rotalock" connector
   nuts ....65-70 ft.-lbs.
                           88.1-94.9 N·m
 Cap screws (hex, 12 point &
   socket head) ....20-24 ft.-lbs.
                           27.1-32.5 N·m
YORK (2 Cylinder)
 "Rotalock" connector
   nuts ....30-35 ft.-lbs.
                           40.7-47.5 N·m
 Cap screws (hex head)....8-13 ft.-lbs.
                           10.8-17.6 N·m

Fig. 6-10—Purge compressor by opening low pressure manifold valve and permitting refrigerant to escape from high pressure service hose connection.

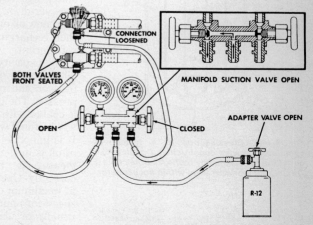

Paragraphs 65-67

AIR CONDITIONING

On models **with manual service valves,** attach the manifold gage set and close both manifold valves and connect the center hose to the Refrigerant-12 container. Loosen the service hose at the high pressure service valve. Open the refrigerant container valve, then slowly open the low pressure valve on manifold for approximately 15 seconds. Allow the refrigerant to sweep through the compressor and out the loosened manifold hose connection. Tighten the loose connection, then close the manifold suction and refrigerant container valves. Turn both service valves counter-clockwise as far as possible.

On models **with automatic Schrader valves** in service ports, refer to preceding paragraph 61 for installation of SWEEP-TEST CHARGE, then evacuate system.

On all models, check refrigerant and add necessary amount of R-12. Install all protective caps when finished.

CHECKING OIL

65. The special refrigeration oil is a highly refined mineral oil with impurities such as wax, moisture and sulphur removed. All refrigerant oil is not the same so be sure to follow manufacturer's recommendations. Viscosity of refrigerant oil is determined by the time in seconds required for a specified quantity of oil to flow through a certain size orifice with oil at 100 degrees F. (37.8 degrees C.). Some of the more commonly used viscosity grades for air conditioners are 300, 500 and 525.

The refrigerant oil is soluable in the refrigerant and consequently is carried throughout the system with the R-12. Refrigerant leakage is often noticed by the telltale oil stain and many potential leaks (such as the compressor front seal) are stopped by oil filling the small gaps.

The oil does not provide cooling and in-fact displaces some refrigerant within the system; however, oil is necessary to lubricate the compressor and, on models so equipped, the expansion valve. Cooling can be affected by too much oil, but damage is sure to result from too little. So maintaining the correct quantity is very important. Approximately three fluid ounces (89 mL) of oil is contained in the evaporator, one fluid ounce (30 mL) is contained in condenser, one fluid ounce (30 mL) is in the receiver-drier and some additional oil remains in the connecting hoses and lines. The oil is circulated with the refrigerant of all systems, but some compressors are equipped with an oil reservoir. Compressors without an oil reservoir, such as Delco Air R-4, the lightweight GM DA-6 and GM Harrison 5 cylinder compressors, must be constantly lubricated by oil entering with the low pressure refrigerant. Be especially careful to assure system has enough oil and is initially charged through high pressure side (without running engine) as outlined in paragraph 63.

Oil should always be checked, on all models if a major loss of oil is suspected, such as results from any large refrigerant leak, including a broken hose or leaking compressor seal. Damage to system components and installation of new (dry) components will also result in loss of oil. Contamination of the oil with water metal particles or any other foreign material will require removing old oil thoroughly cleaning (sweep-charging) system, installing new receiver-drier filling with correct quantity of new oil evacuating (drying) system, then recharging with correct amount of refrigerant.

The capacity and method of checking oil level will depend upon type of compressor and method of mounting compressor. The inside of every component is coated with oil and many components will be filled with various volumes of oil. It is important that the correct amount of oil is in the system and to accurately measure it, the oil must be distributed correctly. Operating the air conditioning at maximum cooling at high idle engine speed for approximately 15 minutes will usually circulate oil so that the level can be checked at the compressor. Refer to the appropriate following paragraphs.

NOTE: It is important that nearly the correct amount of oil be in the system before checking. When installing any new component, attempt to add the same amount of oil which was removed, then check the oil level.

Delco Air R-4 (4 Radial Cylinders)

66. The only lubricant for this compressor is circulated with the refrigerant and the compressor does not have a separate reservoir. Lubricant is permitted to accumulate in limited quantities in the accumulator, then is metered out through the small hole (5 – Fig. 6-10A) to lubricate the compressor. **It is important to have enough lubricant in the system or the compressor will be damaged severely.** The only way to accurately check the volume of lubricant contained in this system is by removing all of the lubricant, then filling the system with the proper amount. When installing a new component, an appropriate amount of lubricant can be installed to compensate for lubricant removed with old part. If a new compressor is installed, add one fluid ounce (30 mL) of 525 viscosity refrigerant oil. If evaporator is renewed, new evaporator should contain three fluid ounces (90 mL) of oil. If condenser is renewed, new condenser should contain one fluid ounce (30 mL) of oil. The accumulator should contain approximately three fluid ounces (90 mL) of oil.

Lubricant can be checked, after discharging the system as described in paragraph 60, then checking the oil contained in the accumulator. All of the oil cannot be drained from the accumulator, but approximately three fluid ounces (90 mL) of oil should pour from removed unit. Add the amount of oil drained or 90 mL, whichever quantity is larger, then evacuate and fill with refrigerant as outlined in paragraphs 62 and 63. Be sure to follow procedure outlined for installing initial charge so that compressor is adequately lubricated while charging.

Delco Air A-6 (6 Axial Cylinders)

67. The oil level can be checked with compressor installed in system or with compressor removed. The preferred method is to purge all old oil and fill with recommended quantity of oil and refrigerant.

To check oil level of installed compressor, operate air conditioner system for 10 minutes with controls set for max-

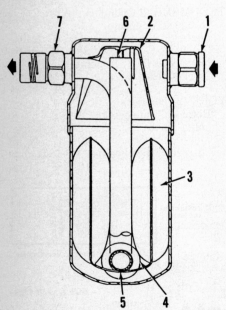

Fig. 6-10A – Cross-section of a typical accumulator showing construction. Opening (6) of outlet tube is covered by baffle (2). Small hole (5) maintains lubricant level within accumulator.

1. Inlet
2. Baffle
3. Desiccant
4. Filter
5. Oil hole
6. Outlet opening
7. Exit connection

SERVICE

Paragraphs 68-69

imum cooling and high blower speed. Engine should be running at about 2000 rpm. Stop engine and momentarily crack compressor drain plug (DP – Fig. 6-11) to let a small amount of oil blow out, then tighten plug. Loosen drain plug slightly for the second time, check for oil, then tighten plug. If oil appears the second time, the compressor has an adequate supply of oil. Foamy oil is considered normal. Absence of oil at second check indicates low oil level. Oil is available in disposable containers which are charged with refrigerant so that oil can be forced into system through normal refrigerant service ports. Application and directions for installation are printed on the containers. Installation is usually accomplished with compressor stopped and can inverted. Be careful not to overfill system. Check refrigerant pressures after servicing with oil. Oil level can also be bench checked as described in the following paragraphs.

To bench check oil level of operating compressor, operate air conditioner for 10 minutes with controls set for maximum cooling and high blower speed. Engine should be running at 1500 rpm. Stop engine, discharge system, then remove compressor. Position compressor horizontal with drain plug downward, remove drain plug and allow all oil to drain from compressor into clean container. Measure the amount of oil removed, then discard old oil. If more than four fluid ounces (118 mL) of old oil was drained from compressor, install same amount of new 525 viscosity oil as was drained. If less than four fluid ounces (118 mL) of oil was drained, add six fluid ounces (177 mL) of Delco #15-117 or equivalent oil, then install drain plug. If oil that was drained contains chips, moisture or other foreign material, flush (sweep-charge) system, install new receiver-drier, evacuate system, install correct total amount (about 11 fluid ounces or 325 mL) of oil, then recharge system with R-12. Be sure to add additional oil if new receiver-drier, evaporator, condenser, etc., is installed. Oil capacity of the entire system is about 11 fluid ounces (325 mL).

Fig. 6-11 – Drain plug (DP) for Delco-Air compressor is used to remove lubricating oil from compressor.

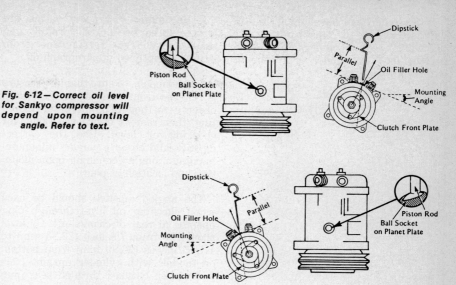

Fig. 6-12 – Correct oil level for Sankyo compressor will depend upon mounting angle. Refer to text.

To bench check oil level of compressor that is not operational (or not installed), remove compressor and drain old oil into clean container. Check quantity and quality of drained oil. The compressor should contain six fluid ounces (177 mL) of oil; however, be sure to add additional oil if other new (dry) components are installed or if system is flushed (sweep-charged). The receiver-drier contains one fluid ounce (30 mL) of oil, the evaporator about three fluid ounces (89 mL) and the condenser about one fluid ounce (30 mL). Additional oil will also coat inside of hoses, etc. Total capacity will be approximately 11 fluid ounces (325 mL). If old oil that was drained from compressor is contaminated with metal particles, moisture or other foreign material, flush system by sweep-charging system, install new receiver drier, evacuate system, install correct total amount of oil, then recharge system with R-12.

Sankyo-Abacus (5 Cylinders)

68. Before checking oil level, operate air conditioner for 10 minutes with controls set for maximum cooling and engine running at idle speed. Release refrigerant from compressor by discharging system as described in paragraph 60 or if possible by isolating compressor, then discharging pressure from compressor as outlined in paragraph 64. Measure the mounting angle of compressor using a Sankyo angle gage (part number 32448) or equivalent. Mounting angle is measured across flats of two front mounting ears and is compared to horizontal. Remove plug from oil filler hole and turn the front clutch plate until internal parts are positioned to permit entrance of the dipstick. Notice the difference depending upon whether compressor is angled to the right or left. Insert the special dipstick (Sankyo part number 32447 or equivalent) into oil filler hole until angle at top of dipstick is flush and parallel with machined boss of oil filler hole. Withdraw dipstick and check to see how many marks are covered by oil. Refer to the accompanying chart and compare actual oil level with acceptable oil level listed for the mounting angle. Only Suniso 5, Texaco Capella "E" or equivalent 500 grade refrigerant oil should be used. Be sure that "O" ring around filler plug is in good condition and not twisted, then install plug to 6-9 ft.-lbs. (8.1-12.2 N·m) torque. Recharge system with refrigerant after oil level is correct and filler plug is tightened. Refer to paragraphs 62 and 63.

Model SD505 compressors contain four fluid ounces (118 mL), Model SD507 compressors contain five fluid ounces (148 mL) and Model SD-508 compressors contain six fluid ounces (177 mL) of oil when system is correctly filled. Amount of oil contained in other parts will vary depending upon size, etc.; but, the evaporator usually contains three three fluid ounces (89 mL), condenser about one fluid ounce (30 mL), receiver-drier about one fluid ounce (30 mL) and connecting lines (hoses) contain about one fluid ounce (30 mL) for each seven feet in length. Too much oil in the system will affect cooling, but insufficient oil will greatly accelerate wear.

Tecumseh (2 Cylinders)

69. Check compressor data plate to be sure of manufacturer. Compressor housing (crankcase) is cast iron. If possible, operate air conditioner for sufficient time to warm compressor before checking oil level.

Paragraph 69 Cont.

AIR CONDITIONING

If the compressor has manual shut-off valves, refer to the appropriate preceding section and isolate compressor from remainder of system as outlined in paragraph 64.

If automatic (Schrader) valves are located in service ports, carefully and safely discharge refrigerant from system as outlined in paragraph 60.

On all models, be sure that all pressure is relieved from compressor crankcase, observe method of compressor mounting, then remove appropriate plug for checking oil level. Use the correct dipstick and observe correct minimum-maximum limits depending upon mounting and location of plug.

The angled dipstick should be used when checking oil level through plug opening at rear of crankshaft. Oil level should be from 7/8 inch (22.23 mm) to 1-1/16 inches (26.99 mm) from bottom on models with cylinder upright (vertical) when checked through rear plug opening. For models with cylinder horizontal, oil level should be within 7/8 inch (22.23 mm) to 1-1/8 inch (28.58 mm) from bottom when checked through rear plug opening.

The curved dipstick should be used when checking oil level through plug opening on side of crankcase. Turn crankshaft to position that will permit easy insertion of dipstick. On models with vertical cylinder, dipstick should be inserted over crankshaft until bottomed in crankcase and permissible oil level is 17/32 inch (13.49 mm) to 1-1/8 inches (28.58 mm). On models with cylinder mounted horizontal, the dipstick should be inserted under crankshaft until bottomed in crankcase and oil level should be 1-15/32 inches (37.31 mm) to 2-9/32 inches (57.94 mm) from bottom of dipstick. The angle of insertion and curvature of dipstick are important when checking oil level. Be sure that end of dipstick is against lowest surface and into corner of crankcase when checking.

Oil capacity of HG850 and HG1000 compressors is approximately 11-12 fluid ounces (325-355 mL), but oil level should be checked as described with dipstick after system has run to be sure

Fig. 6-13—View of cast iron Tecumseh compressor.

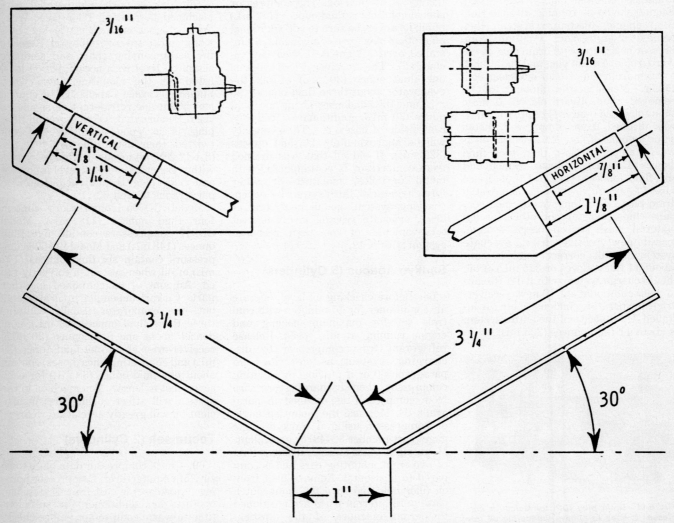

Fig. 6-14—Drawing of dipstick necessary for checking oil level in Tecumseh compressor through opening at rear of crankshaft.

SERVICE

Paragraph 70

of correct amount of oil in complete system. Use only Suniso 5, Texaco Capella "E" or equivalent 500 grade refrigerant oil.

If equipped with manual service valves, purge compressor as described in paragraph 64 before operating system.

If equipped with automatic valves in service ports, sweep-charge, evacuate then fill system with refrigerant as outlined in paragraphs 62 and 63 before operating.

York (2 Cylinders)

70. Check compressor data plate to be sure of manufacturer. Compressor housing (crankcase) is cast aluminum with cast iron cylinder sleeves. If possible, operate air conditioner for sufficient time to warm compressor before checking oil level.

If the compressor is equipped with manual shut-off valves, refer to paragraph 64 and isolate compressor from remainder of system.

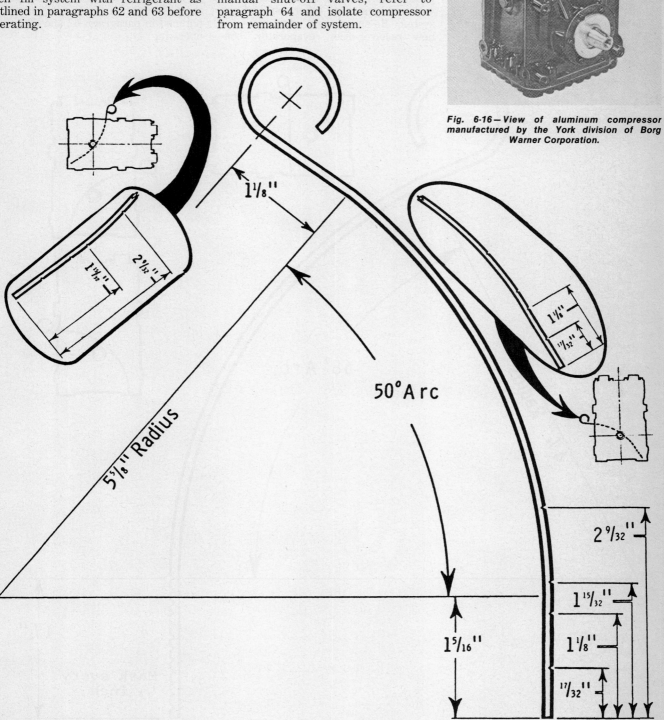

Fig. 6-16 — View of aluminum compressor manufactured by the York division of Borg Warner Corporation.

Fig. 6-15 — Drawing of dipstick suggested for checking oil level in Tecumseh compressor through opening in side of compressor.

Paragraph 70 Cont.

AIR CONDITIONING

If automatic (Schrader) valves are located in service ports, carefully and safely discharge refrigerant from system as outlined in paragraph 60.

On all models, be sure that pressure is relieved from compressor crankcase, observe method of compressor mounting, then remove oil level plug from appropriate location.

A straight dipstick can be used to check oil level of models installed with cylinder horizontal. A curved dipstick must be used on models with cylinder vertical. On all models, turn crankshaft until dipstick can be inserted to bottom of crankcase. Make sure that dipstick is not on rib or boss cast into crankcase or bottom cover when checking. Oil level should be within limits of 7/8 to 1-3/8 inches (22-35 mm) from bottom of crankcase. About one inch (25.4 mm) of oil is approximately eight fluid ounces (237 mL) of oil and is sufficient for compressor lubrication, but oil is also carried with refrigerant to other parts of the system. Be sure to add additional oil if new receiver-drier, evaporator, condenser, etc., is installed. Usual initial oil charge for installing new compressor is 11-12 fluid ounces (325-355 mL). Use only Suniso 5, Texaco Capella "E" or equivalent 500 grade refrigerant oil.

If equipped with manual service valves, purge compressor as described in paragraph 64 before operating system.

If equipped with automatic valves in service ports, sweep-charge, evacuate then fill system with refrigerant as outlined in paragraph 62 and 63 before operating.

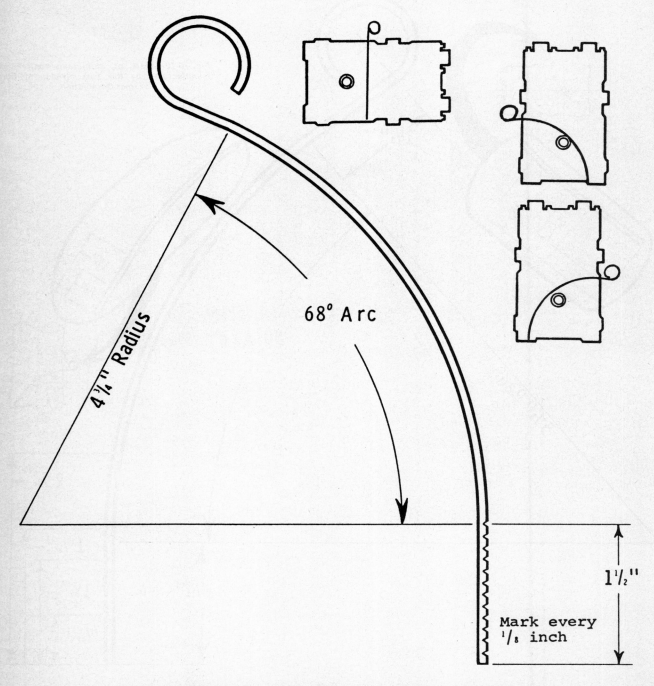

Fig. 6-17 — Drawing of dipstick suggested for checking oil level in York compressor.

SERVICE

Paragraphs 79-82

REPAIR

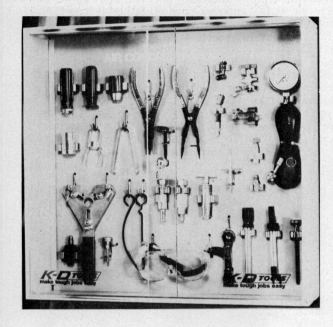

Fig. 7-1 — View of typical tools necessary for minor servicing of air conditioner system. Tools shown are available from K-D Manufacturing Company.

The Nuday Company
2291 Elliott St.
Troy, Michigan 48084

K-D Manufacturing Company
3575 Hempland Rd.
Lancaster, Pennsylvania 17604

Delco Air A-6 (6 Axial Cylinders)

81. The compressor drive clutch (coil, pulley, pulley bearing, drive plate, etc.) can be removed and reinstalled without releasing refrigerant from the system. It will usually be necessary to remove compressor from mounting bracket, but lines may be left connected.

The shaft (front) seal and the high pressure relief valve can be removed and renewed without additional disassembly after refrigerant is discharged from compressor.

82. **R&R COMPRESSOR AND CLUTCH ASSEMBLY.** Discharge refrigerant from system as described in paragraph 60. Disconnect refrigerant lines, then cap all lines and openings in compressor to prevent entry of dirt or other foreign material. Disconnect wire to clutch, then unbolt and remove compressor from mount.

To reinstall, be sure to tighten drive belt sufficiently. Attach hoses and clutch wire, then refer to paragraph 47 to connect manifold and gages to system. Purge compressor as outlined in paragraph 64 if equipped with manual valves. If not equipped with manual type service valves, refer to paragraph 61 to install SWEEP-TEST CHARGE. On all

79. A corrective action should be chosen after determining the cause of any trouble. Often several different repair actions must be used to correct a problem. The most common option is whether to repair the existing faulty unit (or assembly) or to install a new (or rebuilt) similar unit. Be sure to evaluate possible options in light of working conditions (a field is not the most desirable location to overhaul a compressor) as well as length of down time (a new expansion valve can be installed faster and more successfully than repairing the old one.)

CLUTCH AND COMPRESSOR

80. The clutch, front seal, reed valves, etc., can sometimes be removed, serviced (or renewed) and installed with compressor installed. Refer to the appropriate following paragraphs for service. Most service to clutch and compressor will require special tools available from several sources including:

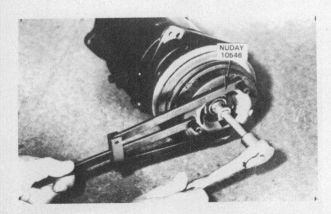

Fig. 7-2 — Holding fixture such as type shown from Nuday Company is useful in holding drive plate.

Fig. 7-3 — A special puller such as type shown from Nuday Company is necessary to remove drive plate without damaging parts.

models the system should be evacuated, then filled with specified amount of oil and refrigerant as described in paragraph 67.

83. R&R CLUTCH ASSEMBLY. Use holding fixture (such as Nuday 10546) and remove locknut, then use puller (Nuday 10504) to remove hub and drive plate from shaft. Examine drive surfaces and renew parts if necessary. If surfaces are damaged, check for cause such as: Low voltage to the engagement coil, binding compressor or refrigerant overcharge. The incorrect coil or a damaged coil may also result in damage to the clutch engaging surfaces.

To remove pulley and bearing, first remove bearing front seal absorbent sleeve, pulley and bearing retaining snap ring and drive key. Use puller (Nuday 10501) to pull bearing and pulley from hub. The bearing can be removed by pressing (or driving) out toward front of bore. Removal and installation of pulley bearing is facilitated using special drivers (Nuday 10502 or equivalent). Install new bearing from front of pulley, pressing on outer race only.

Mark position of coil terminals relative to compressor body. Remove retaining snap ring and lift coil from compressor body. Check coil for loose connections or cracked insulation. Current draw should be approximately 3.2 amps at 12 volts DC at room temperature. Resistance should be approximately 3.66-4.04 ohms.

When assembling, position gasket on compressor, then locate coil being sure that electric terminals are correctly positioned. Install coil retaining snap ring and position felt seal over snap ring.

Push the pulley and bearing onto compressor by carefully pressing only on inner race of bearing. Check pulley for freedom of rotation after pressing into position. Install retaining snap ring with chamfered side out. Install absorbent sleeve with ends butted together toward top of compressor. Install forward seal seat using special tool (Nuday 10478), then install forward dust seal using special tool (Nuday 10438 or equivalent).

Position key in keyway of drive plate so that about 3/8-inch (1 cm) projects

Fig. 7-4 — Pry dust cover from pulley as shown.

Fig. 7-5 — The absorbent sleeve must be in good condition.

Fig. 7-6 — Snap ring shown retains pulley and bearing to hub.

Fig. 7-7 — Special puller such as Nuday tool shown is necessary to remove pulley and bearing. Tool must push on bearing hub, NOT drive shaft.

Fig. 7-8 — Partially disassembled view of typical clutch assembly used on Delco Air compressor. Special nut is shown at (27), drive plate at (28), key at (29), snap ring at (30), pulley and bearing at (31) and coil at (32).

Fig. 7-9 — Special tools such as Nuday tool shown should be used to prevent damage while removing and installing bearing in pulley.

SERVICE

Paragraph 83 Cont.

Fig. 7-10—Special tools such as Nuday models shown will assist in correct positioning of front (dust) seal.

Fig. 7-12—Use the special installing tool to install drive plate on shaft until clearance between drive plate and pulley is about 2 mm (3/32 inch). Nuday No. 10505 tool is shown.

from flange. Align key with keyway in shaft and start hub of drive plate onto shaft. Use special tool (Nuday 10505) to push drive plate onto shaft until clearance between friction surfaces of pulley and drive plate are about 3/32-inch (2 mm) apart.

Two different types of special retaining nuts have been used. The later flanged nut can be used on early models after removing the snap ring and spacer from hub bore. A new nut should be used each time the unit is assembled, regardless of type of nut.

Before installing early type nut, check to be sure that spacer and snap ring are installed in hub. Convex side of snap ring should be in (toward spacer) when correctly positioned. Install new locknut with round machined shoulder toward inside.

On all models, tighten the new special nut to 14-26 ft.-lbs. (19-35 N·m) torque then check clearance between drive plate and pulley friction surfaces. Clearance should be 0.022-0.040 inch (0.56-1.02 mm). Too little clearance may be caused by internal wear and compressor overhaul may be necessary. Excessive clearance may be caused by damaged drive plate.

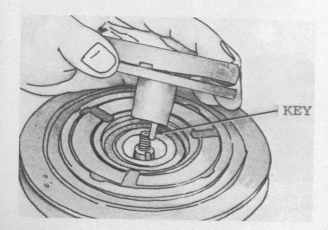

Fig. 7-11—Carefully align key with keyway when installing drive plate. The front (dust) seal is removed for clarity.

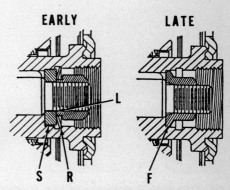

Fig. 7-13—Two types of special nuts have been used. If the late type is installed, be sure that snap ring and spacer are removed.

Paragraph 84

AIR CONDITIONING

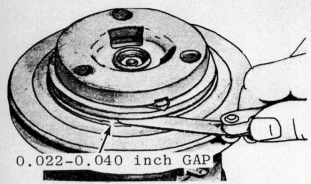

Fig. 7-14—With drive plate correctly installed, clearance between drive plate and pulley will be 0.022-0.040 inch as shown.

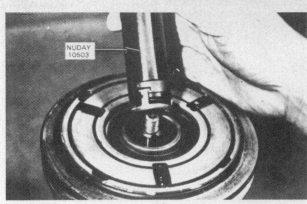

Fig. 7-17—Special holding tools such as Nuday type shown should be used to remove and install seal without damaging seal or other parts.

84. RENEW COMPRESSOR SHAFT FRONT SEAL. Discharge refrigerant from the system as described in paragraph 60. Use a holding fixture (such as Nuday 10546) and remove locknut, then use puller (Nuday 10504) to remove hub and drive plate from shaft. Remove bearing front seal and the absorbent sleeve. Remove seal seat retaining snap ring, insert and engage special seat removing tool (such as Nuday 10513), then withdraw the ceramic seat straight out of bore. Insert special seal removal tool (such as Nuday 10503) into seal bore, engage tool tabs with seal, then withdraw seal from bore. Remove and discard "O" ring from compressor. Inspect compressor bore and shaft to be sure that area is perfectly clean and free from all nicks and burrs.

Lubricate all parts thoroughly with new refrigeration oil before assembling and use only new parts. Lubricate "O" ring and position in special installing tool (Nuday 10509). Install "O" ring in groove of compressor using the special

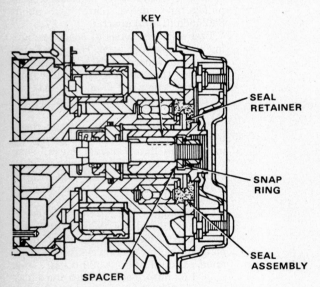

Fig. 7-15—Cross-section of compressor clutch used on Delco Air models.

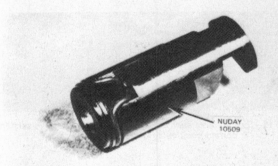

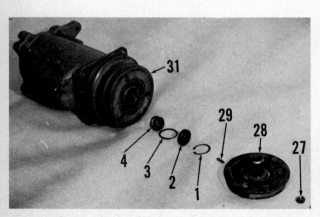

Fig. 7-16—Partially disassembled view of Delco Air compressor showing front seal removed. Seal retaining snap ring is shown at (1), ceramic seat at (2), "O" ring at (3), seal at (4), special nut at (27), and drive plate at (28). Pulley (31) can remain installed.

Fig. 7-18—"O" ring installed in counterbore is more easily installed without damage using one of the special tools available such as the Nuday type shown.

SERVICE

Paragraph 85

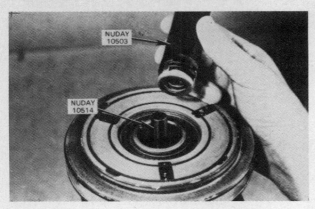

Fig. 7-19—End of shaft must be protected to prevent damage to internal "O" ring in seal assembly when installing new parts. Nuday tools are shown.

Fig. 7-20—Special seal protector should remain in place while installing ceramic seat. Special tool such as Nuday type shown can be used to handle ceramic seat while assembling.

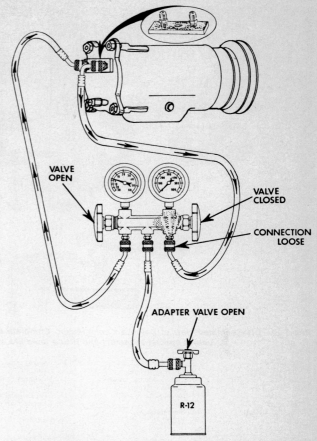

Fig. 7-22—Special test plate such as K-D tool number 2050 can be used to bench check a removed compressor for leaks such as after renewing front seal. Be sure to bleed compressor as shown before checking to remove air from compressor.

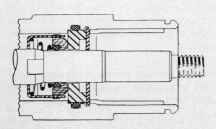

Fig. 7-21—Drawing showing cross-section of seal for Delco Air compressor.

tool. Coat all surfaces of new shaft seal assembly with refrigeration oil, engage seal onto tangs of special installing tool (Nuday 10503) and position seal protector (Nuday 10514) over end of shaft. Slide the seal down into position over shaft applying slight pressure and turning clockwise until seal engages flats on shaft and is firmly seated. Turn the special tool counter-clockwise to disengage from seal, then withdraw tool leaving seal in position. Coat face of new seal seat with refrigerant oil and grip seal seat with special installing tool (Nuday 10513). Push seal into position while rotating so that "O" ring in case bore is not damaged, then remove installation tool. Install new seal retaining snap ring with flat face against seal seat. Tapered side of snap ring must be out. Press snap ring down into bore until it snaps into place in housing groove. Install absorbent sleeve into compressor with ends butted together toward top of compressor (away from oil reservoir side of compressor which must be down when installed). Install forward seal seat using special tool (Nuday 10478), then install forward dust seal using special tool (Nuday 10438). Installation should be checked by following procedure outlined in following paragraph 85, LEAK TEST COMPRESSOR, before proceeding further.

Position key in keyway of drive plate so that about 3/8-inch projects from flange. Align key with keyway in shaft and start hub of drive plate onto shaft. Use special tool (Nuday 10505) to push drive plate onto shaft until clearance between friction surfaces of pulley and drive plate are about 3/32-inch (2 mm) apart. Remove special tool and check to be sure that spacer and snap ring are installed in hub. Convex side of snap ring should be in (toward spacer) when correctly positioned. Install new locknut with round machined shoulder toward inside. Tighten special nut to 14-26 ft.-lbs. (19-35 N·m) torque, then check clearance between drive plate and pulley friction surfaces. Clearance should be 0.022-0.040 inch (0.56-1.02 mm). Clearance of less than 0.022 inch (0.56 mm) may be caused by internal wear and compressor overhaul may be necessary. Clearance of more than 0.040 inch (1.02 mm) may be caused by damaged drive plate.

85. LEAK TEST COMPRESSOR. Oil is necessary for correct sealing especially at the compressor shaft front seal. Add oil if necessary through hole for drain plug in side of compressor housing. Standard oil charge for new compressor is 10-11 fluid ounces (296-325 mL). Install "O" rings and test plate (K-D 2050) to rear of compressor and connect manifold low and high pressure lines to appropriate (suction and discharge) ports. Attach refrigerant

Paragraph 86   AIR CONDITIONING

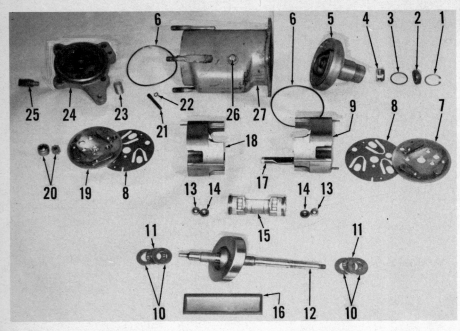

Fig. 7-23 — Disassembled view of Delco Air compressor. Complete disassembly is not suggested unless special assembly and fitting tools are available.

1. Snap ring
2. Ceramic seat
3. "O" ring
4. Front seal
5. Front head
6. "O" rings
7. Front discharge valve plate
8. Suction reeds
9. Front cylinders
10. Thrust bearing races
11. Thrust bearings
12. Swashplate & shaft
13. Balls
14. Shoe discs
15. Piston
16. Suction cross-over cover
17. Discharge cross-over tube
18. Rear cylinders
19. Rear discharge valve plate
20. Oil pump
21. Oil lintake tube
22. "O" ring
23. Screen
24. Rear head
25. Relief valve
26. Drain plug
27. Case

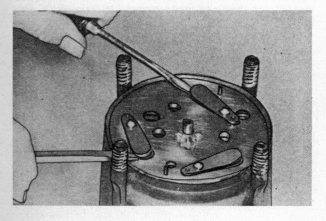

Fig. 7-24 — Two screwdrivers can be used to remove rear discharge valve plate.

Fig. 7-25 — Carefully remove suction valve plate.

container to center (service) hose, loosen high pressure hose connection at manifold, purge lines and compressor, then tighten high pressure connection at manifold. Leave low pressure valve open to permit refrigerant storage pressure of full (or near full) container to pressurize the compressor. Check for leaks especially around compressor shaft front seal and around compressor rear head. Rotate compressor shaft several times while checking for leaks.

86. **COMPRESSOR OVERHAUL.** Internal compressor service requires additional special tools and extreme care must be exercised to assure cleanliness. Many of the internal components establish wear patterns and should not be interchanged with similar but different used parts from the same or another compressor.

To disassemble compressor, first remove the compressor and clutch assembly, then remove clutch and front seal as outlined in preceding paragraphs. Remove compressor drain plug, catch, measure and record amount of oil removed from compressor. Support compressor with drive (front) end down, being careful not to rest any weight on compressor drive shaft. Remove the four nuts securing rear head to compressor, then carefully remove rear head from compressor body. Remove and discard large "O" ring sealing rear head. Lift oil pump inner and outer rotors from shaft and be careful to keep sides of rotors facing rear head together. Most pump rotors have dot marks to indicate side that should be toward rear head. Use two screw drivers and carefully lift rear valve plate from compressor. Remove the suction valve reed disc if not already removed with valve plate. Use needle nose pliers and withdraw oil inlet tube. Push compressor drive shaft by hand out toward rear and withdraw the cylinders and shaft assembly. Remove the front valve plate and suction valve reed.

Examine internal components for damage. Major damage such as caused by operation with insufficient oil is easier and better corrected by installing complete cylinders and shaft assembly. Before disassembling any further, carefully mark all cylinders and pistons for correct assembly into exact same position. A pencil or ink is suggested for marking.

NOTE: Do not bump compressor drive shaft in any way because shaft can be easily bent or otherwise damaged.

Disassemble pistons keeping the parts for each in separate containers. Do not interchange thrust bearings and races if

SERVICE

Paragraph 86 Cont.

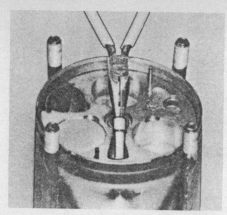

Fig. 7-26—Oil intake tube must be withdrawn before cylinder assembly can be withdrawn from case.

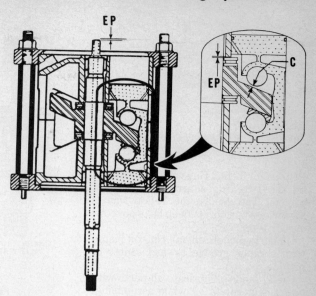

Fig. 7-28—Refer to text for checking shaft end play (EP) and shoe disc clearance (C).

old parts are to be reinstalled. It may be necessary to bend discharge cross-over tube slightly in order to remove shaft. The discharge cross-over tube at original manufacture has ends swedged into place and may be difficult to remove. Service tubes have "O" ring and bushing at each end and can be easily withdrawn by hand. Clean, inspect and air dry all parts.

The shoe discs for pistons are available in eleven different thicknesses to adjust the fit to swashplate to zero clearance. Number stamped on new shoe is thickness in thousandths, such as "18½" is 0.0185 inch thick from surface of shoe to deepest part of ball socket. Correct thickness of shoe can be determined as follows: Assemble pistons, swashplate and cylinders with three shoe discs marked "O" installed only on front end of pistons. Rear end of pistons should not have any shoe installed while checking. Clamp cylinders in test fixture to hold cylinder together, then use spring scale and feeler gage to measure clearance between rear ball and swashplate. Four to eight ounces of pull will be required to withdraw correct thickness of feeler gage from between rear ball and swashplate. Record correct thickness required at that position, turn shaft 120 degrees, check and record correct clearance for that position, then check and record correct clearance for remaining position. Install correct thickness shoe disc for each piston during final assembly.

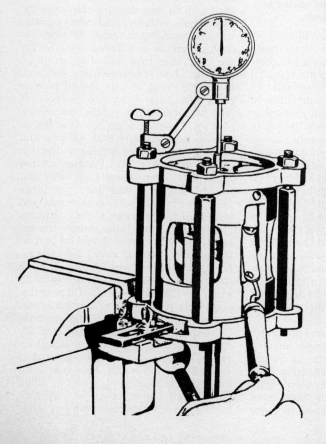

Fig. 7-27—Refer to text for measuring and adjusting shaft end play and shoe disc clearance. Special clamping fixture is required to hold cylinders together while checking.

The thrust races for both front and rear thrust bearings are available in sixteen different thicknesses to adjust shaft end play to zero. Number stamped on thrust race is thickness in thousandths above standard "O" size. Correct thickness of races can be determined as follows: Assemble swashplate in cylinders with four thrust washers marked "O". Clamp cylinders in test fixture to hold cylinders together, then attach dial indicator to check shaft end play. Move shaft up and down and record shaft movement. Select a thrust race 0.0010-0.0015 inch larger than end gap measured. As an example, if end play is 0.0052 inch, remove one of the races marked "O" and install race marked "65" which is approximately 0.0065 inch thicker than the one marked "O".

Pistons for compressors manufactured after 1975 are equipped with Teflon rings which must be cut to size after assembly to the piston. Special tools are available for cutting and checking piston ring size. New pistons are usually supplied with rings presized, but size should be checked before installing.

AIR CONDITIONING

Paragraph 86 Cont.

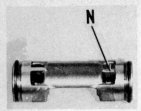

Fig. 7-29 — Forward end of piston has notch at (N).

Pistons for Teflon rings can be identified by the two ridges around bottom of groove. The groove for Teflon ring is also wider but not as deep as groove for earlier type.

On early models, install ring on piston with scraper groove toward center of piston.

On all models, lubricate all parts with new refrigeration oil while assembling. Assemble piston rings to pistons. Install the compressor shaft and swashplate in front housing with threaded end of shaft protruding and forward thrust bearing and races in position. Petroleum jelly can be used to hold bearings in place while assembling. Position balls into ball sockets of one piston and hold the previously selected shoe discs against balls. Identify the forward end of piston by the notch in outside of skirt. Turn swashplate until high point is over the bore for cylinder previously assembled with balls and shoe discs, then insert piston into bore. On early models, ring gap should be toward center (shaft) of compressor. On all models, be sure that notched forward position is inserted into forward cylinder bore. Assemble and install the other two pistons over the swashplate and into correct cylinder bores using same procedure as for the first piston. All pistons and cylinders should have been marked before or during disassembly and should be reinstalled in original positions. Use the shoe discs previously selected to provide zero clearance fit around swashplate.

Install the discharge cross-over tube into hole in front cylinder half, with flattened section toward center (shaft) or compressor.

NOTE: Do not attempt to install original discharge cross-over tube, but instead use service tube which can be seated with "O" rings and bushings at ends.

Rotate shaft to position pistons at three different heights. On early units be sure that rings on pistons are positioned with gaps toward center (shaft), then push rings toward outside of pistons so that gaps will start in bore first. On all models, position rear cylinder half over pistons, then carefully work rings into cylinder bores beginning with the highest piston and continuing to the lowest. After pistons are completely into bores in rear cylinder, align end of cross-over tube with hole in rear cylinder half. Be sure that flat side of tube is toward center of compressor to provide clearance for swashplate. After parts are properly aligned, bump rear cylinder to seat firmly onto locating dowel pins. Install the suction cross-over cover, bending slightly if necessary. Install "O" rings and bushings over ends of the discharge cross-over tube.

Position one of the suction reed plates over shaft and front of cylinder aligning holes for dowel pins, ports, oil return slot and discharge cross-over tube. Position the front discharge valve plate (which has larger center hole) over shaft, aligning holes for dowel pins and

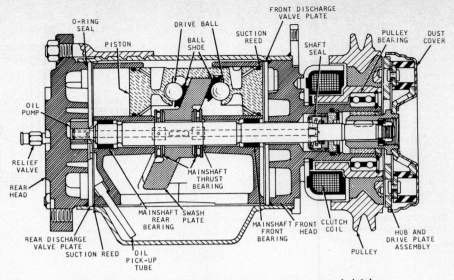

Fig. 7-31 — Cross-section of typical Delco Air compressor and clutch.

for suction valve plate. The discharge reed stops must be away from the previously installed suction reeds. Carefully determine alignment of dowel pins and appropriate holes, then carefully install front head over the shaft, front valves and cylinder assembly. Use care to assure straight alignment, while assembling to prevent damage to valves and other parts caused by turning and misalignment. Lubricate large "O" ring with oil and position in bevel at rear of front head. Coat inside of compressor shell with oil then slide the shell over the front head, cylinders and related parts, carefully working the large "O" ring into shell. Align the hole in rear cylinder for the oil intake tube with bottom of oil sump as shell is lowered into position. Install "O" ring in hole for oil pick-up tube, then slide tube into position. It may be necessary to rotate compressor slightly in shell to align holes for the oil pick-up tube. Be sure that "O" ring and bushing are in place at rear of the discharge cross-over tube. Position suction reed plate over dowel pins with slot toward sump. Install rear discharge plate over dowel pins with reeds and reed stop toward rear. Install oil pump rotors over shaft with marks toward rear. The outer rotor should be pushed toward right so that cavity between inner and outer rotors is formed at 3 o'clock position as viewed from rear with sump at bottom (6 o'clock). Oil rear discharge plate, valves and pump generously, then position large "O" ring into bore of compressor shell against the rear discharge plate. Install suction screen in rear head, coat rear head inner surfaces with oil, then carefully slide rear head into position being careful to align dowels and pump rotor. Be sure that suction screen does not drop out when

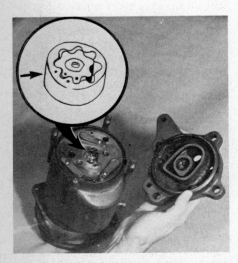

Fig. 7-30 — Push outer rotor of oil pump in direction of arrow before installing rear head.

SERVICE

Paragraphs 87-89

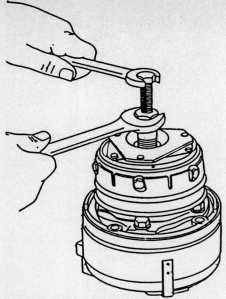

Fig. 7-34 — A special puller can be used as shown to remove the clutch drive plate and hub.

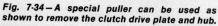

Fig. 7-35 — A special puller and protecting cover for shaft is required when pulling clutch rotor and bearing from front head.

installing. If rear head does not slide down completely, twist front head slightly by hand until rear head slides over dowel pins. Tighten the four nuts retaining rear head to 20 ft.-lbs. (27 N·m) torque for early compressor with iron piston rings; or to 25 ft.-lbs. (34 N·m) for late models with Teflon piston rings. Install shaft front seal and compressor clutch, then leak test as outlined in previous paragraphs. Initial oil charge for compressor is six fluid ounces (177 mL). Only 525 viscosity refrigeration oil should be used.

Delco Air R-4 (4 Radial Cylinders)

87. The compressor contains very little lubricant, but depends upon constant flow of oil within the system, circulated with the refrigerant. It is important that the system always contain an adequate supply of oil or the compressor will be damaged.

The compressor drive clutch (coil, pulley, pulley bearing, drive plate, etc.) can be removed and reinstalled without releasing refrigerant from the system. It may be necessary to remove the compressor from the mounting bracket, but lines may remain connected.

The front seal and the high pressure relief valve can be removed and renewed without additional disassembly after discharging refrigerant from the system.

88. **R&R COMPRESSOR AND CLUTCH ASSEMBLY.** Discharge refrigerant from the system as described in paragraph 60. Disconnect refrigerant lines from compressor, then cover all openings in compressor and lines to prevent the entry of dirt or other foreign material. Disconnect wire to clutch, then unbolt and remove compressor from mount. If excessive leakage is not evident, add one fluid ounce (30 mL) of oil before installing the compressor. If excessive oil leakage is evident or if mechanical damage to the compressor has contaminated the system, more oil is required. Clean all removable filters and install new drier-filter-accumulator assembly if contamination is suspected. The accumulator should contain three fluid ounces (90 mL) of 525 viscosity refrigerant oil, the condenser should contain one fluid ounce (30 mL), and the evaporator contains one fluid ounce (30 mL) of oil and additional oil is necessary to coat the lines and other components.

To reinstall, be sure to tighten drive belt sufficiently. Attach hoses and clutch wire, then refer to paragraph 47 to connect manifold and gages to system. Refer to paragraph 61 to install SWEEP-TEST CHARGE, then evacuate the system as described in paragraph 62. The refrigerant does not contain oil, so be sure that complete system contains the correct amount of oil (paragraph 66), then install initial refrigerant charge (paragraph 63) before operating and filling system.

89. **R&R CLUTCH ASSEMBLY.** Use a suitable holding fixture to keep the hub from turning, then remove the shaft nut. The removed nut should be discarded and a new special nut installed when assembling. Use a suitable puller to remove the hub and drive plate from the compressor shaft.

Examine the clutch drive surfaces and renew parts if necessary. If surfaces are damaged, check for cause such as: Low voltage to the engagement coil, binding compressor, refrigerant overcharge or incorrect engagement coil.

To remove the clutch rotor, pulley and bearing, first remove the retaining snap ring and mark the location of the engagement coil terminals. Use a suitable puller to withdraw the rotor and bearing assembly from the front bearing head. The coil is also pressed into position and will be withdrawn at the same time, unless the pulley is unbolted from the rotor. Remove attaching screws, then separate the rotor, pulley and coil. The screws and lock plates should not be reused.

If only the bearing is to be renewed, the pulley attaching screws can be removed before withdrawing the rotor. The coil will remain in place.

The bearing is pressed and staked into hub of rotor. The bearing can be pulled or pressed from hub without removing the staking metal which retains the bearing. Press new bearing firmly into bore, then use center punch to stake metal at three equally spaced locations around hub. Punch marks should be approximately 1/16-inch (1 mm) deep and should retain bearing, but should not distort bearing outer race.

The pulley and coil must be properly located before installing rotor and bearing. The coil should be installed with terminals correctly positioned and with the three protrusions located in the holes in the front head.

Special tools are available to protect the center of the rotor while driving the rotor and bearing onto the front bearing head. After fully seating rotor and bearing, install the retaining snap ring. Apply "Loctite" or equivalent to the threads of the screws which attach the pulley groove to the rotor. Install screws and special lockwashers, but do not tighten screws until it is certain that the pulley is aligned and rotates freely. Tighten pulley retaining screws to 100 in.-lbs. (11.30 N·m) torque, then bend lock plates around flats of screws.

Shaft and clutch drive hub should be clean and dry without any nicks or burrs. Slide the drive key into groove of the clutch hub, leaving approximately 3/16-inch (5 mm) of the key protruding from the hub to facilitate alignment. Slide hub onto shaft until clearance between face of rotor and clutch plate is 0.020-0.040 inch (0.51-1.02 mm). Do not damage compressor by driving hub onto shaft. A special threaded puller is available to pull the hub onto shaft. Install a new shaft nut with smaller diameter toward inside, then tighten nut to 10 ft.-lbs. (13.56 N·m) torque.

AIR CONDITIONING

Paragraphs 90-92

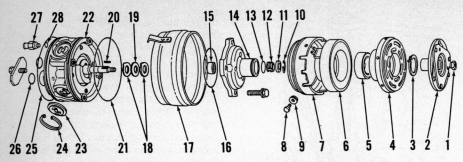

Fig. 7-36 — Exploded view of the four cylinder radial compressor and clutch.

1. Nut
2. Clutch drive plate
3. Snap ring
4. Rotor
5. Bearing
6. Coil
7. Pulley
8. Screw
9. Lockwasher
10. Snap ring
11. Seal seat
12. Seal
13. "O" ring (for seal)
14. Front head
15. Bearing
16. "O" ring seal
17. Shell
18. Thrust washers
19. Belleville washer
20. Drive key
21. "O" ring (front of shell)
22. Cylinder, shaft and pistons assy.
23. Valve plate
24. Snap ring
25. "O" ring (rear of shell)
26. "O" rings (ports)
27. Pressure relief valve
28. "O" ring (relief valve)

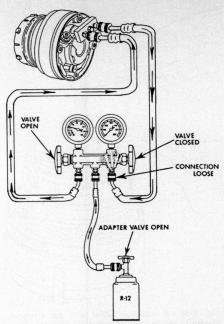

Fig. 7-37 — The compressor can be pressure tested for leakage using a special plate fitted over ports as shown and unit pressurized by refrigerant.

90. RENEW COMPRESSOR SHAFT FRONT SEAL. Discharge refrigerant from the system as described in paragraph 60. Use a holding fixture and remove nut from end of shaft, then use appropriate puller to withdraw hub and drive plate from shaft. Remove the shaft seal retaining snap ring, then thoroughly clean inside of neck bore, front of shaft seal seat and the shaft. Use appropriate seal protector over end of shaft and special seal removing tool withdraw seal seat, seal and "O" ring. Reclean front of compressor neck.

Lubricate all parts thoroughly with new 525 viscosity refrigerant oil before assembling and install only new parts. Lubricate "O" ring and position in special installing tool. Install "O" ring in groove of compressor using the special tool, then coat all surfaces of new shaft seal assembly with refrigeration oil, engage seal onto tangs of special installing tool and position seal protector over end of shaft. Slide the seal down into position over shaft applying slight pressure and turning clockwise until seal engages flats on shaft and is firmly seated. Turn special tool counter-clockwise to disengage from seal, then withdraw tool leaving seal in position. Coat surface of new seal seat with oil and grip seat with special installing tool. Install the seal seat being careful not to dislodge the "O" ring, then disengage and remove the special tool. Install the snap ring with the flat side toward the seal seat.

Leakage at the seal is usually noticed by the presence of lubricant and this oil should be replaced. The most satisfactory method of adding the correct amount of oil is accomplished by removing the accumulator. Carefully drain oil from the accumulator, noticing the amount and checking for contamination. If the oil is contaminated, the accumulator-drier-filter should be renewed and all removable filters should be cleaned. The accumulator should contain at least three fluid ounces (90 mL) of 525 viscosity oil. Fill accumulator with three fluid ounces (90 mL) or the amount drained whichever is the larger amount, then install the accumulator.

NOTE: Compressors manufactured after January 1984 are equipped with a Teflon lip type seal. The later type seal should not be installed on earlier compressors which were originally equipped with the two piece ceramic and carbon seal, because the shaft is not polished smooth enough. Shaft size is the same for both early and later compressors and the two piece seal may be used on all models. If the Teflon lip seal is renewed on later compressors, be sure to protect seal from damage during installation using the appropriate seal protector. Use extreme care, because any roughness of seal protector can easily damage new seal while installing.

The new seal should be checked for leakage before completing the assembly. Pressurize compressor through the low pressure port after blocking the high pressure port. Rotate the compressor shaft by hand several times and check for leakage.

If the front seal is not leaking, remove excess oil and install clutch hub and plate. Insert the drive key part way into keyway of hub leaving about 3/16-inch (5 mm) of the key protruding to facilitate assembly. Do not damage the compressor by driving hub onto shaft. A special threaded puller is available to pull hub onto shaft. Clearance between face of rotor and face of clutch plate should be 0.020-0.040 inch (0.51-1.02 mm). Install a new shaft nut and tighten to 10 ft.-lbs. (13.56 N·m) torque. Small diameter of nut should be toward inside.

91. LEAK TEST COMPRESSOR. Oil is required for complete sealing, especially at the compressor shaft front seal. The compressor should contain approximately one fluid ounce (30 mL) of 525 refrigeration oil. Install "O" rings and test plate to rear of compressor. Connect manifold low and high pressure lines to appropriate (suction and discharge) ports. Attach refrigerant container to center (service) hose, loosen high pressure hose connection at manifold, purge lines and compressor, then tighten high pressure connection at manifold. Leave low pressure valve open to permit refrigerant storage pressure of full (or near full) container to pressurize the compressor. Check for leaks especially around compressor shaft front seal and location around "O" ring seals. Rotate compressor shaft several times while checking for leaks around shaft seal.

92. COMPRESSOR FRONT HEAD. To remove the front head, first discharge refrigerant from the system as described in paragraph 60, then remove the clutch assembly and seal as described in previous paragraphs. Remove the four screws which retain the front head to the compressor, then lift front head from shaft and compressor. Remove "O" ring seal, the two thrust washers and the Belleville washer. The needle bearing can be driven from bore toward rear if renewal is necessary.

Special tool is available to facilitate installation of bearing to correct depth.

SERVICE

Paragraphs 93-96

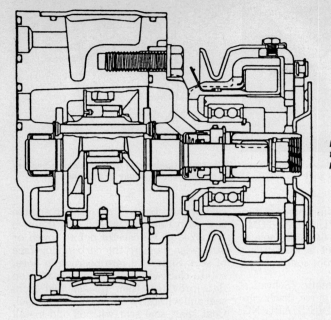

Fig. 7-38 — Cross section of four cylinder radial compressor. Refer to Fig. 7-36 for exploded view.

Lubricate bearing before assembling. Be sure that no dirt, including lint from cleaning cloth, is permitted to remain inside compressor. Coat thrust washers with refrigeration lubricant, then position on shaft. The innermost thrust washer should have tang toward outside, Belleville washer should be installed with high center section out, then outer thrust washer should be installed with tang toward inside. Tangs on both thrust washers should be over the Belleville washer located between them.

Coat "O" ring with lubricant and position in groove of front head. Locate oil hole in front head and assemble so that hole is toward top of compressor when installed. The oil hole delivers lubricant to the bearing and front seal. Tighten the four screws which attach front head to 20 ft.-lbs. (27 N·m) torque. Be sure that "O" ring remains in position when assembling. Refer to paragraphs 89 and 90 for assembling the seal and compressor drive clutch.

93. COMPRESSOR OVERHAUL. Internal compressor service is usually limited to renewal of seals and valves. Extreme care must be exercised to assure cleanliness. Because of the special tools required, it is not recommended to disassemble the cylinders, pistons and shaft assembly.

The compressor shell can be pressed forward off housing after moving the restraining strap out away from rear of compressor housing. The shell is a tight fit and compressor or shell may be damaged if not moved evenly off toward front. Remove and discard old "O" rings and be sure that compressor and shell are completely clean. The discharge valves can be withdrawn after removing the retaining snap rings.

Coat the "O" rings in lubricant, then position the "O" rings in grooves of the housing. Be very careful not to nick or cut the "O" rings while installing or leakage is sure to result. Slide the shell over the housing from the front as far as possible by hand, then use a suitable tool to pull the shell over the lubricated "O" rings. Little force is required if the shell is pulled on straight. Bend the restraining strap around the rear. Be sure to pressure check the compressor after assembly is complete.

Sankyo-Abacus (5 Cylinders)

94. The compressor drive clutch (coil, pulley, pulley bearing, drive plate, etc.) can be removed and reinstalled without releasing refrigerant from the system. It is often necessary (or easier) to remove the compressor from mounting bracket, but lines may be left connected.

Refrigerant must be discharged before front seal or reed valves can be removed.

95. R&R COMPRESSOR AND CLUTCH ASSEMBLY. If possible, isolate compressor as outlined in ISOLATING AND PURGING COMPRESSOR paragraph 64. If not equipped with manual valves to isolate compressor, refer to DISCHARGING SYSTEM paragraph 60 and remove refrigerant from system. After all pressure is removed from compressor, disconnect refrigerant lines, then cap all lines and openings to prevent entry of dirt or other foreign material. Disconnect wire to clutch, then unbolt and remove compressor from mount.

To reinstall, be sure to tighten drive belt sufficiently to prevent slipping. Attach hoses and clutch wire, then refer to ATTACHING MANIFOLD AND LINES paragraph 47 to connect manifold and gages to system. Purge system as outlined in SWEEP-TEST CHARGE paragraph 61. Evacuate system, then fill with specified amounts of oil and refrigerant as described in paragraphs 62 and 63.

96. R&R CLUTCH ASSEMBLY. Use holding fixture to prevent front drive plate from turning and remove shaft front nut with ¾-inch hex socket. Use three leg puller to pull front drive plate from shaft. The drive plate has three ¼-inch-20 threaded holes to which puller can be attached. Remove snap ring holding bearing inner race to hub and use a suitable puller to pull the pulley (with bearings) from compressor hub. Do not press against shaft when

Fig. 7-42 — View of manual service valve (SV) installed on Sanyko compressor.

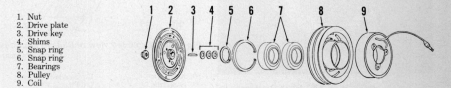

1. Nut
2. Drive plate
3. Drive key
4. Shims
5. Snap ring
6. Snap ring
7. Bearings
8. Pulley
9. Coil

Fig. 7-43 — Compressor clutch typical of type installed on Sanyko Compressor.

AIR CONDITIONING

Paragraphs 97-98

Fig. 7-44—Snap ring shown at (10—Fig. 7-48) must be removed to remove seal.

Fig. 7-46—Special puller will permit easy removal of seal seat.

Fig. 7-49—Tighten rear head retaining screws to recommended torque in sequence shown.

removing pulley, but instead use adapter which pushes against hub. Puller should grip belt groove in circumference of pulley or snap ring groove in inner (bearing) bore of pulley. The clutch energizing coil can be withdrawn after removing the three #10-24 retaining screws. Bearings can be pressed from pulley if service is required.

Install the clutch energizing coil, support compressor by the four rear mounting tabs, align pulley squarely on front hub and press pulley with bearings firmly onto hub. Check to be sure that pulley rotates freely and install retaining snap ring.

Be sure that shims and drive key are installed on shaft, align keyway in drive plate with key, then bump pulley onto shaft. Be careful not to hit shaft. Tighten retaining nut to 25-30 ft.-lbs. (33.9-40.67 N·m) torque. Measure air gap between pulley and drive plate engaging surfaces with feeler gages. If gap is not 0.016-0.031 inch (0.406-0.787 mm) remove nut and clutch drive plate, then add or remove shims as necessary to provide the correct gap. Reinstall and measure gap as previously described after thickness of shims has been changed.

97. RENEW COMPRESSOR SHAFT FRONT SEAL. If equipped with manual type servicing valves, refer to ISOLATING AND PURGING COMPRESSOR paragraph 64 and isolate compressor from rest of system, then discharge refrigerant from compressor.

If equipped with automatic (Schrader type) valves in servicing ports, discharge system as described in DISCHARGING SYSTEM paragraph 60. On all models, use holding fixture to prevent drive plate from turning and remove nut from front of shaft using ¾-inch hex socket. Use three leg puller and three ¼-inch-20 threaded holes in drive plate hub to pull the drive plate from shaft. Remove key and shims from shaft then carefully clean the area around shaft and front of compressor. Remove snap ring from seal bore, then use special seal seat removal/installation tool (Sankyo No. 32405) to remove the seal seat. Use special seal removal/installation tool (Sankyo No. 32425) to withdraw seal from bore. Remove "O" rings from groove in hub bore and from shaft, then clean and flush bore completely.

Install special protector sleeve (Sankyo No. 32426) over end of shaft, lubricate all parts thoroughly with new refrigeration oil before assembling and use only new parts. Lubricate "O" rings and position in groove of hub and around shaft using special tool (Sankyo No. 32406). Coat all surfaces of new shaft seal with refrigeration oil and engage seal with tangs of special tool (Sankyo No. 32425). Be careful not to touch or otherwise damage the carbon ring face of seal. Slide seal into position in bore, rotate clockwise until seal is seated on flats provided, then release tool from seal and withdraw special installing tool. Coat shaft seal seat with oil and install with special lapped surface against carbon ring of seal. Special tool (Sankyo No. 32405) can be used to assure proper installation of seal seat. Install snap ring retaining seal seat with beveled side in against seal seat. Reinstall shims, key and clutch drive plate as described in previous paragraph 96, R&R CLUTCH ASSEMBLY. Check clutch plate to pulley clearance and adjust if necessary. New seal may indicate a slight leak immediately after installation, but should quickly seat together and form an acceptable seal after a short period of normal operation.

98. SERVICE REAR HEAD AND VALVES. The compressor should be removed as outlined in appropriate previous paragraph. Remove the five screws attaching rear head to compressor body using a 13 mm socket. The rear head and valves can be removed but may be stuck to rear of compressor. A gasket scraper or similar tool can be used to separate rear head and valve plate

Fig. 7-45—Special tools are available to facilitate removal and installation of seal.

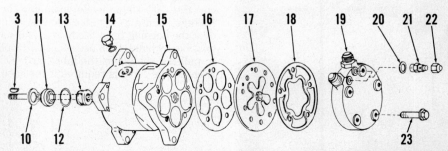

Fig. 7-48—Exploded view of Sankyo compressor reed valve and seal assemblies. Individual parts of compressor (15) are not available.

3. Key
10. Snap ring
11. Seal
12. "O" ring
13. Seal seat
14. Plug
15. Compressor assy.
16. Gasket
17. Reed assy.
18. Gasket
19. Rear head
20. "O" ring
21. Service port
22. Cap
23. Rear head retaining screws

54

SERVICE

Fig. 7-50 — Cross-section of typical Sankyo compressor and clutch assembly.

1. Clutch & pulley
2. Shaft front seal
3. Clutch bearings
4. Thrust bearing
5. Front housing
6. "O" ring
7. Anti-rotation gear
8. Cylinder
9. Piston
10. Cylinder head gasket
11. Reed assy.
12. Rear head
13. Service port cap
14. Hose connection
15. Oil filler plug
16. Planet plate

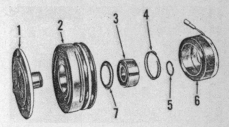

Fig. 7-52 — View of clutch typical of one type used on some Tecumseh compressors.

1. Drive plate
2. Pulley
3. Bearing
4. Snap ring
5. Snap ring
6. Coil
7. Dust cover

from compressor and each other; however, use caution to prevent damaging any of the machined surfaces.

Carefully remove all of the old gasket material, but do not scratch or otherwise mar sealing surfaces. Coat valve cores and "O" rings for service ports with refrigeration oil before installing. Reed valves are available only as an assembly. Check and clean all parts carefully and coat gaskets, sealing surfaces and screws with refrigeration oil before assembly. Position gaskets, valve plate and rear head on rear of compressor. Center of hose connections on rear head must be aligned with oil filler plug on side of compressor body. Tighten the five screws attaching rear head evenly and to a final torque of 22-25 ft.-lbs. (30-34 N·m) using the sequence shown.

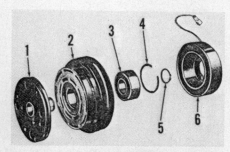

Fig. 7-51 — View of clutch typical of one type used on some Tecumseh compressors.

1. Drive plate
2. Pulley
3. Bearing
4. Snap ring
5. Snap ring
6. Coil

Tecumseh (2 Cylinders)

99. The compressor drive clutch (coil, pulley, bearing and drive plate) can be removed and reinstalled without releasing refrigerant from the system. It is often necessary (or easier) to remove the compressor from mounting bracket, but lines may be left connected.

Refrigerant must be discharged before front seal or reed valves can be removed.

100. **R&R COMPRESSOR AND CLUTCH ASSEMBLY.** If possible isolate compressor as outlined in ISOLATING AND PURGING COMPRESSOR paragraph 64. If equipped with Schrader valves in service ports, refer to DISCHARGING SYSTEM paragraph 60 and remove refrigerant from system. After all pressure is removed from compressor, disconnect refrigerant lines, then cap all lines and openings to prevent entry of dirt or other foreign material. Disconnect wire to clutch, then unbolt and remove compressor from mount.

When reinstalling, tighten mounting screws to 14-17 ft.-lbs. (18.98-23.05 N·m) torque. Tighten drive belt sufficiently to prevent slipping. Attach hoses and clutch wire, then refer to ATTACHING MANIFOLD AND LINES paragraph 47 to connect gages to system. Purge system as outlined in SWEEP-TEST CHARGE paragraph 61. Evacuate system, then fill with sufficient amount of oil and refrigerant as outlined in paragraph 63.

101. **R&R CLUTCH ASSEMBLY.** Remove clutch center screw, then install the correct puller screw into hub of drive plate against compressor shaft. The drive plate will be pulled from tapered end of compressor shaft as puller screw is tightened.

Installation and removal of clutch center screw and puller screw are sometimes easier if clutch is engaged by attaching battery to ground and clutch wire. Holding fixtures are also available to hold drive plate hub.

The clutch field coil can be removed from front of compressor after removing the attaching screws. Clutch coil resistance should be about 2.57-2.83 ohms. Current draw should be 4.45 amperes at 12 volts.

Reinstall by reversing removal procedure. Be sure that clutch ground wire is firmly in contact with clean grounded surface. Tighten drive plate retaining screw to 20 ft.-lbs. (27 N·m) torque. The pulley must spin freely. Install new clutch assembly if parts are bent in such a way to prevent pulley from turning freely.

102. **RENEW COMPRESSOR SHAFT FRONT SEAL.** If equipped with manual type servicing valves, refer to ISOLATING AND PURGING COMPRESSOR paragraph 64 and isolate compressor from rest of system, then discharge refrigerant from compressor. If equipped with automatic (Schrader type) valves in servicing ports, discharge system as described in DISCHARGING SYSTEM paragraph 60. On all models, use holding fixture to prevent drive plate from turning and remove screw from front of shaft using a ½-inch hex socket. Use the correct puller screw threaded into hub of drive plate against compressor shaft to push clutch from tapered end of compressor shaft. Remove Woodruff key and dust shield (if so equipped). Clean all dirt and oil from

AIR CONDITIONING

Paragraph 103

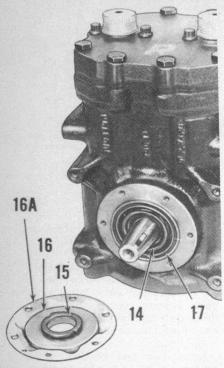

Fig. 7-53 — View of typical Tecumseh compressor with front seal retainer removed. Some differences may be noted between individual units.

14. Seal gland
15. Carbon ring
16. Seal plate
16A. Retainer
17. "O" ring

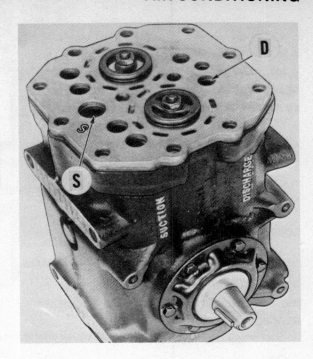

Fig. 7-55 — View of reed plate assembly showing correct installation of suction port (S) and discharge port (D).

area around front of compressor and shaft. Unbolt and remove the seal plate retainer, then withdraw shield, seal plate, "O" ring, carbon ring and seal.

Lubricate all parts of seal, compressor shaft, gaskets and associated parts with refrigeration oil before assembling. Position seal over shaft in bore of compressor with end that contacts carbon ring toward outside (front), then install carbon ring with raised rim out toward front. Insert "O" ring into position in crankcase, then slide seal plate and shield into bore. Install retainer, aligning holes and push against retainer and install retaining screws. Tighten seal plate retainer screws to 54-78 in.-lbs. (6.1-8.8 N·m) torque. Rotate shaft several times by hand to be sure seal is seated. Check oil level as outlined in paragraph 69, then install clutch assembly.

103. RENEW REED PLATE ASSEMBLY. If possible isolate compressor as outlined in ISOLATING AND PURGING COMPRESSOR paragraph 64. If not equipped with manual valves, refer to DISCHARGING SYSTEM paragraph 60 and remove all refrigerant from system. Detach service valves, with hoses attached, from compressor cylinder head. Remove screws attaching cylinder head to compressor, then lift cylinder head and reed valve plate assembly from compressor. Gaskets will probably stick cylinder head, reed plate and cylinder block together and a hammer must be carefully used to separate the pieces.

Clean all of the old gasket material from parts and inspect for damage. Individual reeds are available, but installation of complete reed plate assembly is suggested if reed is damaged. Be sure to check filters and lines for restrictions especially if parts of the reed petals are missing.

Use refrigeration oil on all gaskets before assembly. Do not use sealer which can be squeezed into passages and circulated throughout the system clogging filters and blocking the expansion

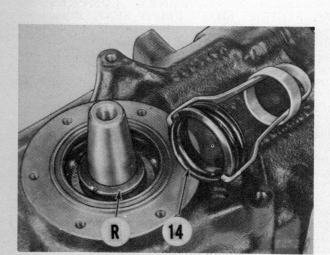

Fig. 7-54 — View of special tool available for removing seal gland (14). Use needle nose pliers to remove rubber seal (R) if it remains on shaft.

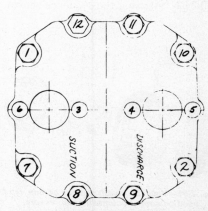

Fig. 7-56 — Tighten cylinder head retaining screws in sequence shown evenly and to the final torque listed in text to prevent leakage.

SERVICE

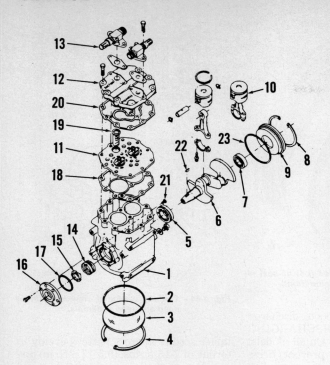

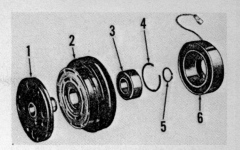

Fig. 7-57 — Exploded view of typical Tecumseh compressor. Several differences may be noted between illustration and actual unit.

1. Crankcase
2. "O" ring
3. Bottom plate
4. Snap ring
5. Front bearing
6. Crankshaft
7. Rear bearing
8. Snap ring
9. Rear bearing housing
10. Connecting rod & piston
11. Reed plate
12. Cylinder head
13. Service valve
14. Seal gland
15. Carbon ring
16. Seal plate
17. "O" ring
18. Gasket
19. Suction inlet screen
20. Gasket
21. Bearing retainer screw
22. Drive key
23. "O" ring

Fig. 7-59 — View of clutch typical of one type used with York compressor. Refer to Fig. 7-60 for legend.

valve orifice. Be sure that gaskets are correctly positioned. Tighten retaining screws evenly until final torque of 20-24 ft.-lbs. (27.12-32.54 N·m) is reached. Service valves attached with Rotalock nuts should have Rotalock nuts tightened to 65-70 ft.-lbs. (88.13-94.91 N·m) torque. Evacuate, purge and refill as outlined in paragraphs 61, 62 and 63 to complete assembly of installed compressor.

104. COMPRESSOR OVERHAUL. The two cylinder reciprocating compressor is serviced in much the same way as a similar sized two cylinder engine. To disassemble, first refer to appropriate paragraphs and remove compressor, clutch, reed plate assembly and front seal. Remainder of disassembly and service will be obvious and depend upon extent of repair needed. Compressors rebuilt by commercial rebuilders or new compressors are often more practical for service than overhauling one used unit.

The front bearing lock screws should be tightened to six ft.-lbs. (8.13 N·m) torque; the connecting rod screws to seven ft.-lbs. (9.49 N·m) torque. Model HG-850 compressor has 1-7/8 inch (47.6 mm) bore, 1-35/64 inch (39.3 mm) stroke and total displacement of the two cylinders is 8.54 cubic inches (139.9 cc). Model HG-1000 compressor has 1-7/8 inch (47.6 mm) bore, 1-7/8 inch (39.3 mm) stroke and total displacement of the two cylinders is 10.35 cubic inches (169.6 cc).

York (2 Cylinders)

105. The compressor drive clutch (coil, pulley, bearing and drive plate) can be removed and reinstalled without releasing refrigerant from the system. It is often necessary (or easier) to remove the compressor from mounting bracket, but lines may be left connected.

Refrigerant must be discharged before front seal or reed valves can be removed.

106. R&R COMPRESSOR AND CLUTCH ASSEMBLY. If possible isolate compressor as outlined in ISOLATING AND PURGING COMPRESSOR paragraph 64. If equipped with Schrader valves in service ports, refer to DISCHARGING SYSTEM paragraph 60 and remove refrigerant from system. After all pressure is removed from compressor, disconnect refrigerant lines, then cap all lines and openings to prevent entry of dirt or other foreign material. Disconnect wire to clutch, then unbolt and remove compressor from mount.

When reinstalling, tighten mounting screws to 14-17 ft.-lbs. (18.98-23.05 N·m) torque. Tighten drive belt sufficiently to prevent slipping. Attach hoses and clutch wire, then refer to ATTACHING MANIFOLD AND LINES paragraph 47 to connect gages to system. Purge system as outlined in SWEEP-TEST CHARGE paragraph 61. Evacuate system, then fill with sufficient amount of oil and refrigerant as described in paragraphs 62 and 63.

107. R&R CLUTCH ASSEMBLY. Remove clutch center screw, then install the correct puller screw into hub of drive plate against compressor shaft. The drive plate will be pulled from tapered end of compressor shaft as puller screw is tightened.

Installation and removal of clutch center screw and puller screw are sometimes easier if clutch is engaged by attaching battery to ground and clutch wire. Holding fixtures are also available to hold drive plate hub.

The clutch field coil can be removed from front of compressor after removing the attaching screws. Clutch coil

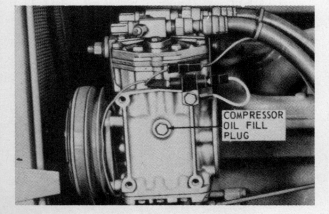

Fig. 7-58 — View of typical York compressor installed. Oil fill plug is usually located in such a way to permit checking oil level without removing compressor. Unit shown has manual service valves.

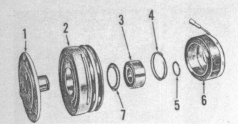

Fig. 7-60—View of clutch typical of one type used with York compressor.

1. Drive plate
2. Pulley
3. Bearing
4. Snap ring
5. Snap ring
6. Coil
7. Dust cover

resistance should be about 2.57-2.83 ohms. Current draw should be 4.45 amperes at 12 volts.

Reinstall by reversing removal procedure. Be sure that clutch ground wire is firmly in contact with clean grounded surface. Tighten drive plate retaining screw to 20 ft.-lbs. (27 N·m) torque. The pulley must spin freely. Install new clutch assembly if parts are bent in such a way to prevent pulley from turning freely.

108. RENEW COMPRESSOR SHAFT FRONT SEAL. If equipped with manual type servicing valves, refer to ISOLATING AND PURGING COMPRESSOR paragraph 64 and isolate compressor from rest of system, then discharge refrigerant from compressor. If equipped with automatic (Schrader

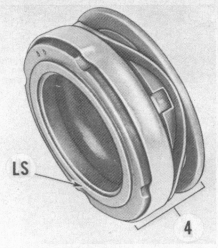

Fig. 7-62—The lapped surface (LS) of seal (4) should be toward outside (front).

type) valves in servicing ports, discharge system as described in DISCHARGING SYSTEM paragraph 60. On all models, use holding fixture to prevent drive plate from turning and remove screw from front of shaft using a ½-inch hex socket. Use the correct puller screw threaded into hub of drive plate against compressor shaft to push clutch from tapered end of compressor shaft, then remove Woodruff key and dust cap (if so equipped). Clean all dirt and oil from area around front of compressor and shaft. Unbolt and remove the seal plate, then withdraw carbon ring, "O" ring, and seal.

Lubricate all parts of seal, compressor shaft, "O" ring and associated parts with refrigeration oil before assembling. Push seal over shaft in bore of compressor with raised, lapped surface (LS–Fig. 7-52) that contacts carbon ring toward outside (front). Locate carbon ring, "O" ring and seal plate over end of shaft, push parts into position and install retaining screws. Tighten re-

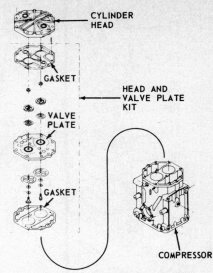

Fig. 7-64—Partially exploded view of cylinder head and reed assembly.

tainer screws in cross pattern evenly to torque of 7-13 ft.-lbs. (9.5-17.6 N·m) torque. Rotate shaft several times by hand to be sure seal is seated. Check oil level as outlined in paragraph 70, then install clutch assembly.

109. RENEW REED PLATE ASSEMBLY. If possible isolate compressor as outlined in ISOLATING AND PURGING COMPRESSOR paragraph 64. If not equipped with manual valves, refer to DISCHARGING SYSTEM paragraph 60 and remove all refrigerant from system. Detach service valves, with hoses attached, from compressor cylinder head. Remove screws attaching cylinder head to compressor, then lift cylinder head and reed valve plate assembly from compressor. Gaskets will probably stick cylinder head, reed plate and cylinder block together and a hammer must be carefully used to separate the pieces.

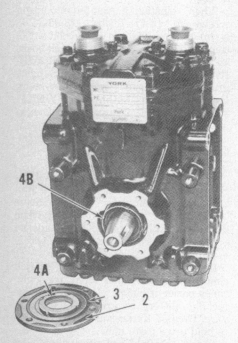

Fig. 7-61—View of York compressor with clutch and seal retainer (2) removed. Carbon ring (4A), seal gland (4B) and "O" ring (3) are also shown.

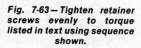

Fig. 7-63—Tighten retainer screws evenly to torque listed in text using sequence shown.

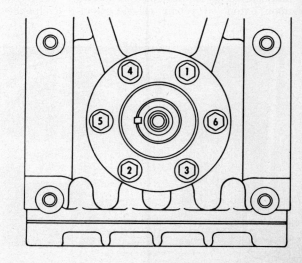

SERVICE

Paragraph 110

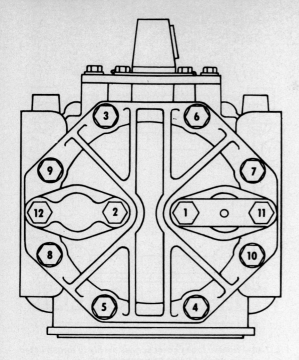

Fig. 7-65—Tighten cylinder head evenly to torque listed in text using sequence shown.

circulated throughout the system clogging filters and blocking the expansion valve orifice. Be sure that gaskets are correctly positioned. Tighten cylinder head retaining screws evenly, in sequence shown, until final torque of 15-23 ft.-lbs. (20.34-31.18 N·m) is reached. Service valves attached with Rotalock nuts should have Rotalock nuts tightened to 30-35 ft.-lbs. (40.67-47.45 N·m) torque. Evacuate, purge and refill as outlined in paragraphs 61, 62 and 63 to complete assembly of installed compressor.

110. **COMPRESSOR OVERHAUL.** The two cylinder reciprocating compressor is serviced in much the same way as a similar sized two cylinder engine. To disassemble, first refer to appropriate paragraphs and remove compressor, clutch, reed plate assembly and front seal. Remainder of disassembly and service will be obvious and depend upon extent of repair needed. Compressors rebuilt by commercial rebuilders or new compressors are often more practical for service than overhauling one used unit.

Clean all of the old gasket material from parts and inspect for damage. Individual reeds are available, but installation of complete reed plate assembly is suggested if reed is damaged. Be sure to check filters and lines especially if parts of the reed petals are missing.

Use refrigeration oil on all gaskets before assembly. Do not use sealer which can be squeezed into passages and

The rear cover plate screws should be tightened to 9-17 ft.-lbs. (12.2-23.0 N·m) torque; the connecting rod screws to

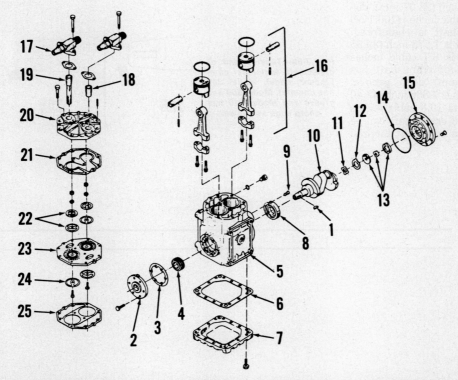

Fig. 7-66—Exploded view of typical York compressor. Several differences may be noticed between various units.

1. Drive key
2. Retainer plate
3. Gasket (or "O" ring)
4. Seal assy.
5. Cylinder
6. Gasket
7. Bottom cover
8. Front bearing
9. Bearing retainer screw
10. Crankshaft
11. Wave washer
12. Thrust washer
13. Oil pump
14. "O" ring
15. Pump housing & rear cover
16. Connecting rod & piston assy.
17. Service valve
18. Discharge tube
19. Suction tube with screen
20. Cylinder head
21. Gasket
22. Discharge reed & retainer
23. Reed plate
24. Suction reed
25. Gasket

59

Paragraph 110 Cont.   AIR CONDITIONING

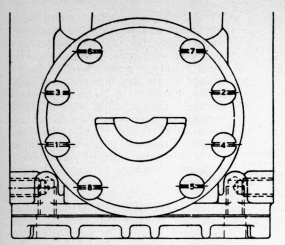

Fig. 7-67—Tighten rear cover plate screws evenly to torque listed in text using sequence shown.

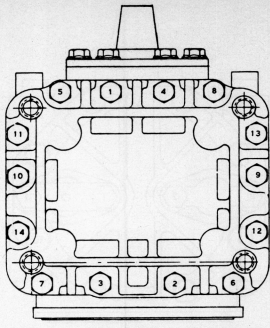

Fig. 7-68—Tighten lower cover screws evenly to torque listed in text using sequence shown.

13-16 ft.-lbs. (17.6-21.7 N·m) torque; the lower cover (base plate) screws should be tightened to 14-22 ft.-lbs. (19.0-29.8 N·m) torque. The compressor model can be identified by the chamfer, groove or sharp corner at front (tapered) end of crankshaft. All 206, 209 and 210 models have 1.875 inch (47.63 mm) bore diameter and two cylinders. Stroke of 206 model is 1.105 inch (28.07 mm), displacement is 6.11 cubic inches (100.1 cc) and end of crankshaft is chamfered. Stroke of 209 model is 1.573 inch (39.95 mm), displacement is 8.7 cubic inches (142.6 cc) and end of crankshaft has 0.015 inch (0.38 mm) deep groove. Stroke of 210 models is 1.866 inch (47.40 mm), displacement is 10.3 cubic inches (168.8 cc) and end of crankshaft can be identified by a sharp corner at tapered end.

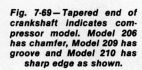

Fig. 7-69—Tapered end of crankshaft indicates compressor model. Model 206 has chamfer, Model 209 has groove and Model 210 has sharp edge as shown.

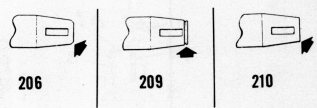

TROUBLESHOOTING

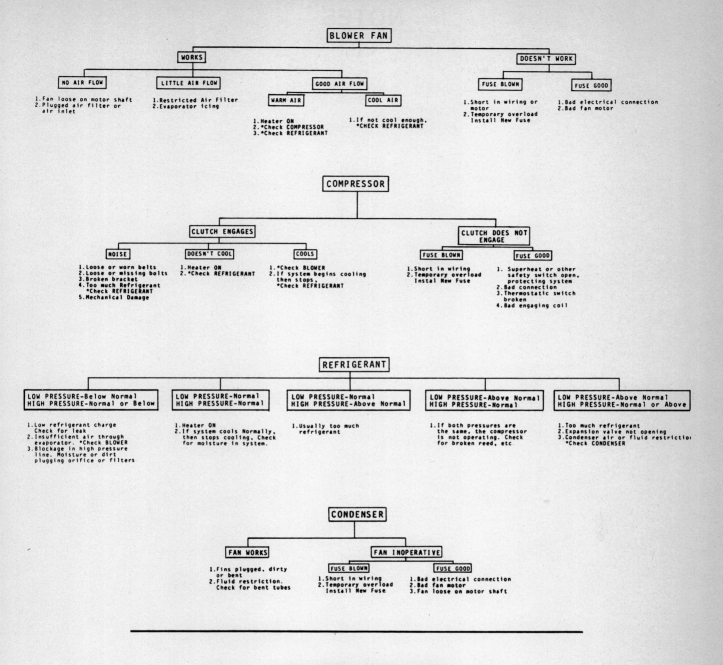

APPROXIMATE SYSTEM PRESSURE

Ambient Temperature – Degrees ...	70 F. (21 C.)	80 F. (27 C.)	90 F. (32 C.)	100 F. (38 C.)	110 F. (43 C.)
High Pressure – Gage Reading	145-155 psi (1000-1069kPa)	170-180 psi (1172-1241kPa)	200-210 psi (1379-1448kPa)	215-225 psi (1482-1551kPa)	220-230 psi (1517-1586kPa)
Low Pressure – Gage Reading	7-15 psi (48-103kPa)	7-15 psi (48-103kPa)	7-15 psi 48-103kPa)	7-30 psi (48-207kPa)	7-35 psi (48-241kPa)
Approximate Evaporator Discharge Air Temperature – Degrees	(7.8-9.5 C.)	8.5-10.0 C.)	(9.5-11.1 C.)	(10.0-11.7 C.)	(10.6-12.2 C.)

NOTES

METRIC CONVERSION

MM.	INCHES			MM.	INCHES			MM.	INCHES			MM.	INCHES			MM.	INCHES			MM.	INCHES		
1	0.0394	1/32	+	51	2.0079	2.0	+	101	3.9764	3 31/32	+	151	5.9449	5 15/16	+	201	7.9134	7 29/32	+	251	9.8819	9 7/8	+
2	0.0787	3/32	−	52	2.0472	2 1/16	−	102	4.0157	4 1/32	−	152	5.9842	5 31/32	+	202	7.9527	7 15/16	+	252	9.9212	9 29/32	+
3	0.1181	1/8	−	53	2.0866	2 3/32	−	103	4.0551	4 1/16	−	153	6.0236	6 1/32	−	203	7.9921	8.0	−	253	9.9606	9 31/32	−
4	0.1575	5/32	+	54	2.1260	2 1/8	+	104	4.0945	4 3/32	+	154	6.0630	6 1/16	+	204	8.0315	8 1/32	+	254	10.0000	10.0	
5	0.1969	3/16	+	55	2.1654	2 5/32	+	105	4.1339	4 1/8	+	155	6.1024	6 3/32	+	205	8.0709	8 1/16	+	255	10.0393	10 1/32	+
6	0.2362	1/4	−	56	2.2047	2 7/32	−	106	4.1732	4 3/16	−	156	6.1417	6 5/32	−	206	8.1102	8 1/8	−	256	10.0787	10 3/32	−
7	0.2756	9/32	−	57	2.2441	2 1/4	−	107	4.2126	4 7/32	−	157	6.1811	6 3/16	−	207	8.1496	8 5/32	−	257	10.1181	10 1/8	−
8	0.3150	5/16	+	58	2.2835	2 9/32	+	108	4.2520	4 1/4	+	158	6.2205	6 7/32	+	208	8.1890	8 3/16	+	258	10.1575	10 5/32	+
9	0.3543	11/32	+	59	2.3228	2 5/16	+	109	4.2913	4 9/32	+	159	6.2598	6 1/4	+	209	8.2283	8 7/32	+	259	10.1968	10 3/16	+
10	0.3937	13/32	−	60	2.3622	2 3/8	−	110	4.3307	4 11/32	−	160	6.2992	6 5/16	−	210	8.2677	8 9/32	−	260	10.2362	10 1/4	−
11	0.4331	7/16	−	61	2.4016	2 13/32	−	111	4.3701	4 3/8	−	161	6.3386	6 11/32	−	211	8.3071	8 5/16	−	261	10.2756	10 9/32	−
12	0.4724	15/32	+	62	2.4409	2 7/16	+	112	4.4094	4 13/32	+	162	6.3779	6 3/8	+	212	8.3464	8 11/32	+	262	10.3149	10 5/16	+
13	0.5118	1/2	+	63	2.4803	2 15/32	+	113	4.4488	4 7/16	+	163	6.4173	6 13/32	+	213	8.3858	8 3/8	+	263	10.3543	10 11/32	+
14	0.5512	9/16	−	64	2.5197	2 17/32	−	114	4.4882	4 1/2	−	164	6.4567	6 7/16	−	214	8.4252	8 7/16	−	264	10.3937	10 13/32	−
15	0.5906	19/32	−	65	2.5591	2 9/16	−	115	4.5276	4 17/32	−	165	6.4961	6 1/2	−	215	8.4646	8 15/32	−	265	10.4330	10 7/16	−
16	0.6299	5/8	+	66	2.5984	2 19/32	+	116	4.5669	4 9/16	+	166	6.5354	6 17/32	+	216	8.5039	8 1/2	+	266	10.4724	10 15/32	+
17	0.6693	21/32	+	67	2.6378	2 5/8	+	117	4.6063	4 19/32	+	167	6.5748	6 9/16	+	217	8.5433	8 17/32	+	267	10.5118	10 1/2	+
18	0.7087	23/32	−	68	2.6772	2 11/16	−	118	4.6457	4 21/32	−	168	6.6142	6 5/8	−	218	8.5827	8 19/32	−	268	10.5512	10 9/16	−
19	0.7480	3/4	−	69	2.7165	2 23/32	−	119	4.6850	4 11/16	−	169	6.6535	6 21/32	−	219	8.6220	8 5/8	−	269	10.5905	10 19/32	−
20	0.7874	25/32	−	70	2.7559	2 3/4	−	120	4.7244	4 23/32	−	170	6.6929	6 11/16	+	220	8.6614	8 21/32	−	270	10.6299	10 5/8	−
21	0.8268	13/16	+	71	2.7953	2 25/32	+	121	4.7638	4 3/4	+	171	6.7323	6 23/32	+	221	8.7008	8 11/16	+	271	10.6693	10 21/32	+
22	0.8661	7/8	−	72	2.8346	2 27/32	−	122	4.8031	4 13/16	−	172	6.7716	6 25/32	−	222	8.7401	8 3/4	−	272	10.7086	10 23/32	−
23	0.9055	29/32	−	73	2.8740	2 7/8	−	123	4.8425	4 27/32	−	173	6.8110	6 13/16	−	223	8.7795	8 25/32	−	273	10.7480	10 3/4	−
24	0.9449	15/16	+	74	2.9134	2 29/32	+	124	4.8819	4 7/8	+	174	6.8504	6 27/32	+	224	8.8189	8 13/16	+	274	10.7874	10 25/32	+
25	0.9843	31/32	+	75	2.9528	2 15/16	+	125	4.9213	4 29/32	+	175	6.8898	6 7/8	+	225	8.8583	8 27/32	+	275	10.8268	10 13/16	+
26	1.0236	1 1/32	+	76	2.9921	3.0	−	126	4.9606	4 31/32	−	176	6.9291	6 29/32	−	226	8.8976	8 29/32	−	276	10.8661	10 7/8	−
27	1.0630	1 1/16	+	77	3.0315	3 1/32	+	127	5.0000	5.0		177	6.9685	6 31/32	−	227	8.9370	8 15/16	−	277	10.9055	10 29/32	−
28	1.1024	1 3/32	+	78	3.0709	3 1/16	+	128	5.0394	5 1/32	+	178	7.0079	7.0	+	228	8.9764	8 31/32	+	278	10.9449	10 15/16	+
29	1.1417	1 5/32	−	79	3.1102	3 1/8	−	129	5.0787	5 3/32	−	179	7.0472	7 1/16	−	229	9.0157	9 1/32	−	279	10.9842	10 31/32	+
30	1.1811	1 3/16	−	80	3.1496	3 5/32	−	130	5.1181	5 1/8	−	180	7.0866	7 3/32	−	230	9.0551	9 1/16	−	280	11.0236	11 1/32	−
31	1.2205	1 7/32	+	81	3.1890	3 3/16	+	131	5.1575	5 5/32	+	181	7.1260	7 1/8	+	231	9.0945	9 3/32	+	281	11.0630	11 1/16	+
32	1.2598	1 1/4	+	82	3.2283	3 7/32	+	132	5.1968	5 3/16	+	182	7.1653	7 5/32	+	232	9.1338	9 1/8	+	282	11.1023	11 3/32	+
33	1.2992	1 5/16	−	83	3.2677	3 9/32	−	133	5.2362	5 1/4	−	183	7.2047	7 7/32	−	233	9.1732	9 3/16	−	283	11.1417	11 5/32	−
34	1.3386	1 11/32	−	84	3.3071	3 5/16	−	134	5.2756	5 9/32	−	184	7.2441	7 1/4	−	234	9.2126	9 7/32	−	284	11.1811	11 3/16	−
35	1.3780	1 3/8	−	85	3.3465	3 11/32	−	135	5.3150	5 5/16	−	185	7.2835	7 9/32	−	235	9.2520	9 1/4	−	285	11.2204	11 7/32	+
36	1.4173	1 13/32	+	86	3.3858	3 3/8	+	136	5.3543	5 11/32	+	186	7.3228	7 5/16	+	236	9.2913	9 9/32	+	286	11.2598	11 1/4	+
37	1.4567	1 15/32	−	87	3.4252	3 7/16	−	137	5.3937	5 13/32	−	187	7.3622	7 3/8	−	237	9.3307	9 11/32	−	287	11.2992	11 5/16	−
38	1.4961	1 1/2	−	88	3.4646	3 15/32	−	138	5.4331	5 7/16	−	188	7.4016	7 13/32	−	238	9.3701	9 3/8	−	288	11.3386	11 11/32	−
39	1.5354	1 17/32	+	89	3.5039	3 1/2	+	139	5.4724	5 15/32	+	189	7.4409	7 7/16	+	239	9.4094	9 13/32	+	289	11.3779	11 3/8	+
40	1.5748	1 9/16	+	90	3.5433	3 17/32	+	140	5.5118	5 1/2	+	190	7.4803	7 15/32	+	240	9.4488	7 7/16	+	290	11.4173	11 13/32	+
41	1.6142	1 5/8	−	91	3.5827	3 9/16	−	141	5.5512	5 9/16	−	191	7.5197	7 17/32	−	241	9.4882	7 1/2	−	291	11.4567	11 15/32	−
42	1.6535	1 21/32	−	92	3.6220	3 5/8	−	142	5.5905	5 19/32	−	192	7.5590	7 9/16	−	242	9.5275	9 17/32	−	292	11.4960	11 1/2	−
43	1.6929	1 11/16	+	93	3.6614	3 21/32	+	143	5.6299	5 5/8	+	193	7.5984	7 19/32	+	243	9.5669	9 9/16	+	293	11.5354	11 17/32	+
44	1.7323	1 23/32	+	94	3.7008	3 11/16	+	144	5.6693	5 21/32	+	194	7.6378	7 5/8	+	244	9.6063	9 19/32	+	294	11.5748	11 9/16	+
45	1.7717	1 25/32	−	95	3.7402	3 3/4	−	145	5.7087	5 23/32	−	195	7.6772	7 11/16	−	245	9.6457	9 21/32	−	295	11.6142	11 5/8	−
46	1.8110	1 13/16	−	96	3.7795	3 25/32	−	146	5.7480	5 3/4	−	196	7.7165	7 23/32	−	246	9.6850	9 11/16	−	296	11.6535	11 21/32	−
47	1.8504	1 27/32	+	97	3.8189	3 13/16	+	147	5.7874	5 25/32	+	197	7.7559	7 3/4	+	247	9.7244	9 23/32	+	297	11.6929	11 11/16	+
48	1.8898	1 7/8	+	98	3.8583	3 27/32	+	148	5.8268	5 13/16	+	198	7.7953	7 25/32	+	248	9.7638	9 3/4	+	298	11.7323	11 23/32	+
49	1.9291	1 15/16	−	99	3.8976	3 29/32	−	149	5.8661	5 7/8	−	199	7.8346	7 27/32	−	249	9.8031	9 13/16	−	299	11.7716	11 25/32	−
50	1.9685	1 31/32	−	100	3.9370	3 15/16	−	150	5.9055	5 29/32	−	200	7.8740	7 7/8	−	250	9.8425	9 27/32	−	300	11.8110	11 13/16	−

NOTE. The + or − sign indicates that the decimal equivalent is larger or smaller than the fractional equivalent.

NOTES

GLOSSARY

ABSOLUTE ZERO—Total absence of heat.

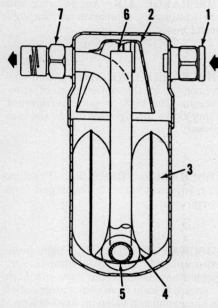

Cross-section of accumulator showing typical construction. Opening (6) of outlet tube is covered by baffle (2). Small hole (5) maintains lubricant level within the accumulator.

1. Inlet
2. Baffle
3. Desiccant
4. Filter
5. Oil hole
6. Outlet opening
7. Exit connection

ACCUMULATOR—A storage device which receives vapor and some remaining liquid refrigerant from the evaporator.

AIR CONDITIONING—Changing temperature, humidity, cleanliness and movement of air.

AIR INLET VALVE—A door that permits entrance of either outside air or inside air to the plenum assembly.

AIR MIX VALVE—A door which opens necessary ducts to blend hot and cold air for desired results. Position is usually controlled automatically by the programmer. Sometimes called blend door.

AIR OUTLET VALVE—A door that directs air through either the heater core or the evaporator.

AMBIENT SENSOR—A thermister mounted to monitor incoming (outside) air temperature.

AMBIENT SWITCH—Switch which prevents operation of compressor below certain (usually 32°F) temperatures.

AMBIENT TEMPERATURE—Temperature of surrounding air. Air entering system before being heated or cooled.

AMPLIFIER—Device used to increase the power of an electrical signal from a sensor. Output from amplifier is then used to control operation of another component such as a transducer.

ASPIRATOR—A suction device used to move air into or across sensor to provide more accurate sample.

A.T.C.—Automatic Temperature Control.

ATMOSPHERIC PRESSURE—Weight of air at a given altitude. Pressure at sea level is approximately 14.7 psi. Pressure decreases at higher altitude.

BI-LEVEL OPERATION—Operation after temperature stabilizes and blower is operating at low speed. Conditioned air exits from both air conditioning and heat outlets.

BOILING POINT—The temperature at which a liquid changes to a vapor.

BRAZING—A metal joining process which is satisfactory for relatively high pressure such as encountered in the condenser.

BTU (British Thermal Unit)—Heat required to raise one pound of water one degree Fahrenheit.

CCOT—Clutch Cycling Orifice type air conditioning system is one design type.

CENTIGRADE—Scale of temperature measurement at which (standard atmospheric pressure) water freezes at 0° and boils at 100°.

CHANGE OF STATE—Rearrangement of the molecular structure as matter changes between two of the physical states (solid, liquid or vapor).

CHARGING—The act of adding refrigerant and/or oil to the air conditioning system.

CHARGING STATION—A single compact unit combining gage set, vacuum pump and refrigerant tank for servicing air conditioning systems.

CHEMICAL INSTABILITY—The undesirable condition sometimes caused by certain contaminants mixed with refrigerant.

CIRCUIT—A path through which electrical current, fluid or a gas can flow from a source, through various units, then back to the source.

CIRCUIT BOARD—A single board containing one or more circuits. Usually electrical circuits, such as used for air conditioning automatic control.

GLOSSARY Continued

CLUTCH — A coupling which controls transfer of movement from a driving member to a driven member when desired.

COLD — Absence of heat.

COMPRESSOR — Pump used to increase pressure of vaporized refrigerant.

CONTROL HEAD — The unit which contains the controls which can be adjusted during normal operation.

CONDENSATION — The process of changing from a gas to liquid. Refrigerant is changed from gas to liquid in the system condenser by increasing the pressure and removing heat from the refrigerant. Water is a product of condensation from the air on the exterior surface of the cold evaporator or a glass of ice in a warm room.

CONDENSING TEMPERATURE — The temperature at which a gas changes to a liquid at a given pressure.

CONDENSING PRESSURE — Pressure indicated by gage attached to service valve located between compressor outlet and condenser inlet.

CONDUCTION OF HEAT — The ability of heat to move between different substances.

CONTAMINANTS — Anything except correct refrigerant and correct oil in a refrigeration system. Moisture, rust, dirt, corrosion and air are most common.

CONVECTION — The movement of heat by circulation of liquid or gas.

CYCLING CLUTCH ORIFICE TUBE (CCOT) — The removable tube containing the fixed orifice.

CYCLING CLUTCH SYSTEM — Cooling temperature is controlled by cycling compressor clutch on, then off to engage.

DARLINGTON AMPLIFIER — A two transister circuit contained in one package. Use on some models as power amplifier for signal to transducer.

DEFROST DOOR — A door which partially blocks or opens a passage which directs heated air toward the windshield. This door is sometimes controlled by a vacuum diaphragm.

DEHYDRATE — Remove moisture or dry out.

DENSITY — Relating to the amount of a substance contained in a given space.

DESICCANT — A substance used to absorb and hold moisture. A dry desiccant is usually contained in a bag located in the receiver or accumulator to absorb and hold moisture from the refrigerant.

DIAGNOSIS — Procedure used to locate the cause of malfunction.

DISCHARGE — To release something which is controlled. Refrigerant should be discharged from system slowly and should be controlled.

DISCHARGE AIR — Air entering cab or passenger compartment through outlet vents.

DISCHARGE LINE — The tube or hose connected to the outlet port of compressor. The high pressure refrigerant vapor flows through this line to the condenser.

DISCHARGE PRESSURE — Pressure of refrigerant which is discharged from compressor.

DISCHARGE SIDE — The high pressure outlet port of compressor. Sometimes includes the complete high pressure system from compressor outlet to the expansion valve or fixed orifice.

DOOR LINK — Linkage connecting programmer output to vent doors.

DRIER — Component of the system containing a desiccant or drying agent. Receivers and accumulators often incorporate a desiccant; however, a drier usually only contains the drying agent and filter.

ELECTRICAL HARNESS — Wires connecting the various electrical components. Usually equipped with multi terminal ends with several wires molded or wrapped together.

ENGINE IDLE COMPENSATOR — On non-diesel engines, a thermostatically controlled system which supplies additional air to idle mixture to prevent stalling during prolonged hot weather operation.

GLOSSARY Continued

ENGINE THERMAL SWITCH — A temperature controlled switch (usually bi-metal) which senses engine temperature and prevents automatic temperature control until engine reaches a specific operating temperature.

EQUALIZER — A line, passage or connection used specifically for operation of certain valves.

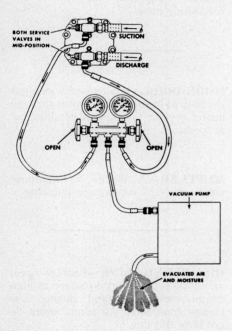

EVACUATE — Create a vacuum within the refrigeration system to lower the boiling point and thus remove moisture (water) from the system. Ambient temperature must be warm enough to cause boiling at the reduced pressure (vacuum) to be effective.

EVAPORATION — Changing liquid to vapor (gas).

EVAPORATOR — Component which cools the air. Liquid refrigerant is changed to vapor (boiled) by absorbing heat from the cab or passenger compartment.

EXPANSION TUBE — The fixed orifice located in the evaporator inlet line of some systems. The expansion tube is removable from most models.

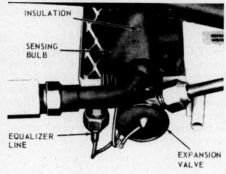

EXPANSION VALVE — A valve located at the entrance of the evaporator which controls the amount of refrigerant permitted to enter the evaporator.

FAHRENHEIT — Scale of temperature measurement at which (standard atmospheric pressure) water freezes at 32° and boils at 212°.

FEEDBACK POTENTIOMETER — A potentiometer which is moved by changing position of a control or vacuum motor.

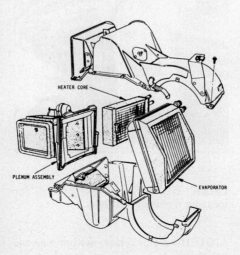

FLUSH — To remove solid material such as dirt and metal particles. Refrigerant passages may be safely flushed (purged or swept) using the standard refrigerant. Other solvents can be used to flush exterior surfaces.

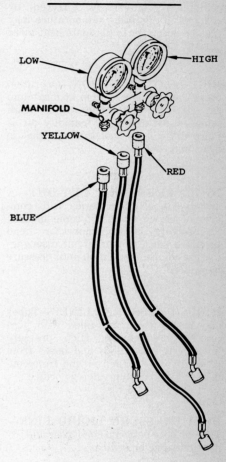

GAGE SET — A system of gages and valves attached to a manifold designed to assist air conditioning measuring system pressures, discharging, evacuating and filling system.

GAS — A vapor which contains no liquid.

HEAD PRESSURE — High (compressed) pressure of refrigerant at output of compressor.

67

GLOSSARY Continued

HEAT EXCHANGE — Movement of heat between two substances of different temperature. Heat always moves from warmer object to colder.

HEAT SINK — An object which absorbs the heat from another object. Electrical components adversely effected by changes in operating temperature may require heat sink to dissipate heat more quickly.

HIGH LOAD CONDITION — Operation of air conditioning system at maximum capacity for extended time. High temperature and high humidity air is more difficult to cool than air that is drier and cooler.

HIGH PRESSURE CUT-OUT — A switch connected in series with compressor clutch circuit of some models. Excessively high compressor head pressure causes open circuit disengaging the compressor clutch until pressure is reduced.

HIGH PRESSURE LINES — Tubes and hoses which are filled with high pressure refrigerant. High pressure lines include all tubes and hoses from compressor to condenser and from condenser to expansion valve or orifice.

HIGH PRESSURE LIQUID LINE — Tube or hose between condenser and expansion valve or orifice.

HIGH PRESSURE RELIEF VALVE — A safety valve located in the low pressure line which can vent a small amount of refrigerant to lower the static pressure to a safe limit. Excessively high pressure can be caused by high ambient temperature and over filled system when compressor is not operating.

HIGH PRESSURE VAPOR LINE — Tube or hose between compressor outlet port and condenser.

HIGH SIDE — The high pressure part of the system from the compressor outlet to the expansion valve or fixed orifice; including condenser, receiver and lines connecting these parts and compressor.

HIGH SIDE PRESSURE — Pressure at test connection on high pressure side of system.

HOT GAS BY-PASS — A line and valve used on some models to prevent moisture from freezing on outside of evaporator, by controlling the temperature of the evaporator. High pressure vapor from compressor outlet is metered by valve to evaporator outlet as needed.

HOT WATER VALVE — A valve which controls the flow of hot water to the heater core. Valve may be opened and closed manually or controlled by vacuum supplied by temperature control.

HUMIDITY — Moisture (water) suspended in the air.

LATENT HEAT — The amount of heat absorbed or released while a substance is changing from one state (gas, liquid or solid) to another.

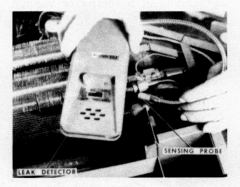

LEAK DETECTOR — Equipment for locating a refrigerant leak.

LIQUID — A fluid state; neither gaseous nor solid.

LOW PRESSURE PROTECTION — A pressure sensitive electrical switch may be connected in series with the compressor clutch electrical control to prevent clutch from operating if refrigerant pressure is less than the switch closing pressure. A safety switch to prevent damage caused by operating without sufficient refrigerant charge.

LOW SIDE — The low pressure part of the system from expansion valve or fixed orifice to the inlet to the compressor, including evaporator, accumulator and connecting lines.

MODE DOOR — A door which changes the path of incoming air to pass through heater core or through air conditioning evaporator.

MUFFLER — A device to minimize noise caused by compressor pumping.

OIL BLEED LINE — Passage providing oil return to compressor at high compressor speed and during low charge conditions. All compressors do not have this line.

OIL INJECTION — Some air conditioner service equipment includes a cylinder for adding a measured amount of refrigerant oil while adding refrigerant.

OPERATIONAL TEST — Run or attempt to run the system usually in all operating modes while checking performance to determine if system is operating satisfactorily or identifying operation failures.

OUTSIDE AIR DOOR — A door permitting outside air into plenum during all modes of operation except recirculation.

GLOSSARY Continued

PHOSGENE—A poisonous gas caused by burning Refrigerant 12 with an open flame.

PLENUM ASSEMBLY—The large chamber which is the central point for all air before passing through the heater or the air conditioning. The plenum contains doors controlling entrance of air from outside or inside and doors which control the exit of air to the heater core or air conditioning evaporator.

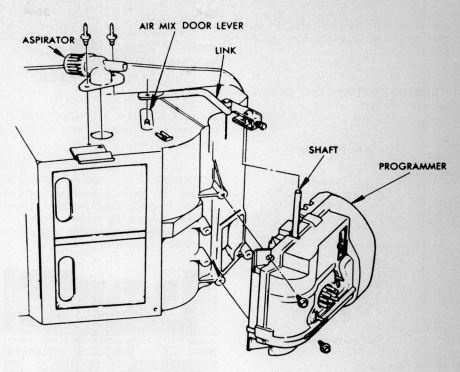

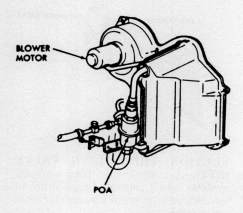

POA—Pilot Operated Absolute. One type of air conditioning control valve.

POWER SERVO—A device used in some systems to control various components which require larger amounts of power to actuate. Usually controlled by a vacuum or low voltage electrical signal.

POWER SPRING—A spring used on some systems to assist operation of a vacuum diaphragm or vacuum motor in one of the two directions.

PRESSURE—Force exerted per unit of area. Standard measurements include pounds per square inch and Neutons per square centimeter.

PRESSURE LINE—All air conditioning refrigerant lines contain pressure; however, the high pressure lines are sometimes referred to as pressure lines.

PROGRAMMER—An assembly used with some systems which coordinates automatic operation of blower speeds, air mix door and vacuum diaphragms in a predetermined sequence.

PSI—Unit of pressure measurement. Pounds per square inch.

PURGE—Flush system or component with clean dry refrigerant to remove moisture, air and contaminants.

RADIATION—The movement of heat through a space.

RECEIVER-DRIER (Receiver-Dehydrator)—A container for storing liquid refrigerant in the line between the condenser and the expansion valve. The container contains a bag of desiccant which can remove small traces of moisture from system.

REFRIGERANT (Refrigerant-12)—The chemical fluid used in automotive air conditioning systems. Freon 12 is a trade name.

REFRIGERANT OIL—Special oil used to lubricate seals, gaskets, compressor and other parts within the air conditioning system. Several different types are available which are especially for-

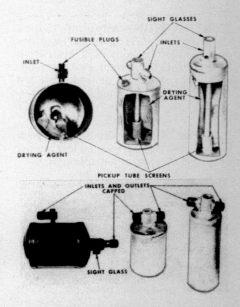

GLOSSARY Continued

mulated to be capable with Refrigerant-12 and lubricate specific systems.

REFRIGERATION — The mechanical removal of heat.

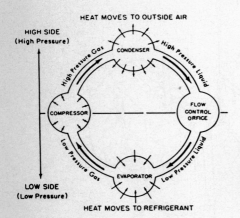

REFRIGERATION CYCLE — The complete flow of refrigerant through the system back to the originating point during normal operation.

RELATIVE HUMIDITY — The moisture content of air in relation to the maximum amount the air can hold at the same temperature. At 75% humidity, the air can only contain ¼ more moisture.

RESTRICTION — Anything which slows the flow of refrigerant. The orifice or expansion is a normal restriction designed into the system. Partially plugged filters or screens or bent tubes are restrictions which should be repaired.

RESTRICTOR — A porous plug which may be installed in a vacuum line to slow operation of a vacuum diaphragm.

SCHRADER VALVE — A spring loaded valve which can be opened by pushing the center rod, such as used in the valve stems of pneumatic tires. Similar but

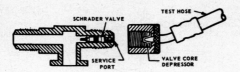

different Schrader valves are used at test point openings of many air conditioning systems.

SCREENS — A fine mesh screen may be located at various locations within system to catch solid particles. Usual locations are at the receiver-drier, expansion valve, accumulator and at entrance to compressor.

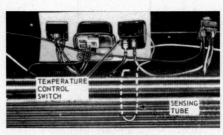

SENSOR — A component which reacts to external changes in a specific way and transmits a resulting signal to a control unit. Sensors may be used to monitor temperature, pressure, humidity or other conditions.

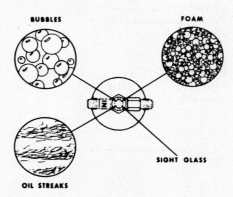

SIGHT GLASS — A small glass window sometimes located in the top of the receiver-drier which can be used to observe the flow of refrigerant, bubbles or air.

SUCTION LINE — The low pressure line which connects evaporator outlet to compressor inlet. Actually, the line should contain low pressure vapor.

SUCTION PRESSURE — Gage pressure measured at inlet of compressor.

SUCTION SIDE — The low pressure part of the system from the expansion valve or orifice to inlet of the compressor. Includes evaporator, accumulator and lines between these parts and compressor.

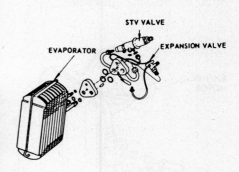

SUCTION THROTTLING VALVE (STV) — A control valve used on some systems which controls temperature of evaporator to prevent ice on exterior of evaporator core.

SUCTION SERVICE VALVE (SSV) — The valve located near the compressor inlet on the low pressure (suction) side of system. The valve is used to attach gage for checking pressure and for charging system.

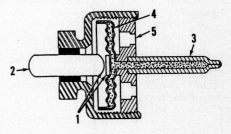

1. Contacts
2. Terminal
3. Sensing tube
4. Diaphragm
5. Openings.

SUPERHEAT SWITCH — Safety switch which opens electrical circuit preventing operation with system low on refrigerant charge. Superheat switches are sometimes used to short a thermo melt fuse by making a ground connection.

GLOSSARY Continued

TEMPERATURE — Measurement of heat intensity.

TEMPERATURE CONTROL — The control which can be moved by the operator to change the automatically adjusted temperature setting.

THERMISTER — An electrical resistor made from material which changes resistance in relation to the temperature. Use as temperature sensor.

TORQUE — Turning force as applied to an object that resists turning.

TRANSDUCER — A device used in some systems to change vacuum output in relation to an electrical input signal. The vacuum output controls movement of vacuum motor or diaphragm.

THERMOSTATIC EXPANSION VALVE (TX) — The valve that controls the amount of liquid refrigerant permitted to enter the evaporator. The amount of refrigerant is determined by the temperature of the evaporator sensor.

VACUUM — Less than atmospheric pressure can be expressed as a vacuum.

VACUUM HARNESS — Vacuum lines which connect various components of the air conditioning control system.

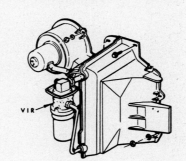

VACUUM MOTOR — A device which converts a vacuum signal to a mechanical action.

VACUUM PROGRAM DISC — A control device used to open or close vacuum passages in a predetermined sequence.

VALVES IN RECEIVER (VIR) — Construction type which incorporates certain valves as part of the receiver-drier. Thermostatic Expansion and Pilot Operated Absolute suction throttling valves are usually those built in.

VAPORIZATION POINT — The temperature at which a liquid changes to a gas.

WIDE OPEN THROTTLE CUT-OFF — A switch which opens the electrical circuit to the air conditioning compressor clutch during full throttle operation, such as during quick acceleration.

NOTES

NOTES

These lined pages have been provided as a handy location for listing often used phone numbers and addresses of suppliers, listing part numbers of common (or unusual) components, keeping service records or specifications and for keeping other important air conditioning service notes.

NOTES

NOTES